FIBRE OPTIC METHODS FOR STRUCTURAL HEALTH MONITORING

FIBRE OPTIC METHODS FOR STRUCTURAL HEALTH MONITORING

Branko Glišić

Smartec SA, Switzerland

Daniele Inaudi

Smartec SA, Switzerland

BICENTENNIAL
1807
WILEY
2007
BICENTENNIAL

John Wiley & Sons, Ltd

Other Wiley Editorial Offices

John Wiley & Sons Inc., 111 River Street, Hoboken, NJ 07030, USA

Jossey-Bass, 989 Market Street, San Francisco, CA 94103-1741, USA

Wiley-VCH Verlag GmbH, Boschstr. 12, D-69469 Weinheim, Germany

John Wiley & Sons Australia Ltd, 42 McDougall Street, Milton, Queensland 4064, Australia

John Wiley & Sons (Asia) Pte Ltd, 2 Clementi Loop #02-01, Jin Xing Distripark, Singapore 129809

John Wiley & Sons Canada Ltd, 6045 Freemont Blvd, Mississauga, ONT, L5R 4J3, Canada

Wiley also publishes its books in a variety of electronic formats. Some content that appears in print may not be
available in electronic books.

Anniversary Logo Design: Richard J. Pacifico

British Library Cataloguing in Publication Data

A catalogue record for this book is available from the British Library

ISBN 978-0470-06142-8

Typeset in 10/12pt Times by Thomson Digital, India
Printed and bound in Great Britain by Antony Rowe Ltd, Chippenham, Wiltshire
This book is printed on acid-free paper responsibly manufactured from sustainable forestry
in which at least two trees are planted for each one used for paper production.

To

Tanja and Lana

and to Morena and Selena

Contents

Foreword

The development of smart structures and structural health monitoring concepts in the civil engineering field has become more and more attractive in the last decade and has received growing attention worldwide in academic and applied research. The basic ideas have been derived from applications performed in the aeronautical, aerospace and automotive industries, but the migration to the civil construction industry has definitely required, and still requires, the development of domain-specific technologies and know-how for the fabrication of sensors, monitoring systems design, data collection and data fusion, analysis and interpretation of the measurements and decision making.

The introduction of fibre-optic sensory systems and related interpretation techniques has contributed to a very significant extent to cover the gap between the above pioneering concepts and practice, thus making possible the realization of extremely reliable monitoring systems that are able to keep under control the behavioural conditions of real structures in all the phases of their existence, from construction to maintenance interventions and practically for their entire operational life.

However, it is observed that, despite these developments, only a limited, although continuously growing, number of practical applications can be reported to date. Two main reasons can be individuated for such a finding. The first reason is that, although observational methods have been the basis for many engineering disciplines, modern structural monitoring techniques are not yet a part of the standard educational programmes of structural engineers and, therefore, they are not well known among most professionals. The second reason is that cost efficiency of structural health monitoring systems in building and infrastructure management can only be demonstrated in the medium to long term.

This book by Branko Glišić and Daniele Inaudi is a significant contribution in overcoming both these difficulties, because it explains with very simple and effective language the most important aspects of selecting, designing and using health monitoring systems based on fibre-optic sensor technologies and presents a wide series of case studies through which the type and quality of the information that can be gathered from these systems is clearly exemplified.

The way in which the different principles and manufacturing techniques are used for the sensors and how these sensors may be placed in structural members to derive local and global behavioural parameters appears to be very suitable for class teaching purposes, but the exhaustive description of the data interpretation approaches and the presentation of the results of several important applications to many different classes of structures will also be of benefit for practising engineers.

Andrea Del Grosso

Professor of Structural Engineering
The University of Genoa, Italy
Genoa, April 2007

Preface

The domain of structural health monitoring has witnessed an impressive development in the last two decades, thanks, on the one hand, to a more widespread acceptance of its benefits by the structure's owners and, on the other hand, to the emergence of new enabling technologies.

Structural health monitoring has found interesting applications in two types of structure in particular: innovative new structures and problematic ageing structures. In the case of newly built constructions, it has become common practice to instrument those that present innovative aspects in terms of the types of material used (e.g. composites or high-performance concrete), structural design or size. On the other hand, old structures with known problems have benefited from structural health monitoring to extend their useful lifespan safely, making full use of the available structural reserves.

On the technology side, new types of sensors and data acquisition systems have appeared, allowing a more reliable and economic instrumentation of many types of structure. Fibre-optic sensors are one of the most prominent technologies that have successfully migrated from the laboratory to the field, and many sensor types have appeared and filled different application niches. In the case of civil structures, the main benefits of fibre optics have been found in their long-term stability and reliability, as well as in their insensitivity to the external perturbations that often affect conventional sensors.

Some of the newly available fibre-optic sensors are the equivalent of existing conventional sensors and can be used as one-to-one replacements of those. For example, this is the case of a point sensor measuring strain or temperature, where the fibre-optic equivalent of a strain gauge or a thermocouple can be used in much the same way. Professionals used to designing, installing and operating electric-based sensor networks can, therefore, migrate to fibre-optic technology with minimal retraining. There are, however, new classes of fibre-optic sensors, in particular of long-gauge and distributed fibre-optic sensors, which have little or no equivalent in the realm of conventional sensing and, therefore, require a different approach.

In the last 15 years we have been fortunate to witness and participate in the development of fibre-optic sensors and their application to structural health monitoring of civil structures. In our activities, however, we observe that a gap still exists between the possibilities offered by modern structural health monitoring technologies and their application in the field. Many practising engineers are not fully aware or convinced by the benefits of applying a monitoring system to their structures and those topics are only marginally covered in the university curricula. In particular, there is a lack of a recognized design methodology for structural health monitoring systems, and many installations are driven by the desire to apply a specific sensing technology rather than selecting the most appropriate solution to a specific monitoring problem. We have also found it difficult to explain the benefits of long-gauge and distributed fibre-optic sensors to instrumentation engineers experienced in the use of point sensors. To realize the full potential of these technologies it is often necessary to approach an instrumentation project from a different angle rather than simply introduce fibre-optic sensors in the same network that would have been used with conventional sensors.

This book was born as an attempt to condense our structural health monitoring methodology into a simple, practical but systematic approach. The concepts and technologies presented in these pages are the result of our own field experience, matured by instrumenting hundreds of structures worldwide, but we do not pretend to cover all existing fibre-optic sensing technologies and their possible application to structural health monitoring. We hope that the readers will be able to apply the methodology presented to their specific monitoring goals and that the many application examples will serve as a field guide to the growing and exciting world of structural health monitoring.

We encourage you to share with us your ideas and comments about this book and the topics presented so that we can make it better and more useful in the future.

Daniele Inaudi (inaudi@smartec.ch)
and **Branko Glišić** (glisic@smartec.ch)

SMARTEC SA, Manno, Switzerland
(www.smartec.ch)
Lugano, 30 April 2007

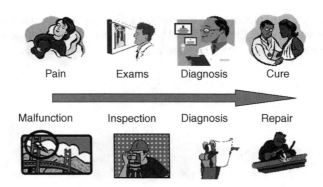

Figure 1.1   Monitoring as structure's feelings (courtesy of SMARTEC).

1. Detect the malfunction in the structure (e.g. crack occurrence, ...)
2. Register the time of problem occurrence (e.g. 19 July 2004 at 14:30, ...)
3. Indicate physical position of the problem (e.g. in the outer beam, 3 m from abutment, ...)
4. Quantify the problem (e.g. open for 2 mm, ...)
5. Execute actions (e.g. turn the red light on and stop the traffic!).

Monitoring is not supposed to make a diagnosis; to make a diagnosis and propose the cure it is necessary to carry out a detailed inspection and related analyses.

Detection of unusual structural behaviours based on monitoring results is performed in accord with predefined algorithms. These algorithms can be simple (e.g. comparison of measured parameters with ultimate values), advanced (e.g. comparison of measured parameters with designed values) or very sophisticated (e.g. using statistic analysis). The efficiency of monitoring depends on both the performance of the applied monitoring system and the algorithms employed. Simple and advanced algorithms are presented in a general manner in Chapter 3. The presentation of sophisticated algorithms exceeds the scope of this book.

1.1.3 Monitoring Needs and Benefits

In the first place, monitoring is naturally linked with safety. Unusual structural behaviours are detected in monitored structures at an early stage; therefore, the risk of sudden collapse is minimized and human lives, nature and goods are preserved.

Early detection of a structural malfunction allows for an in-time refurbishment intervention that involves limited maintenance costs (Radojicic *et al.*, 1999).

Well-maintained structures are more durable, and an increase in durability decreases the direct economic losses (repair, maintenance, reconstruction) and also helps to avoid losses for users that may suffer due to a structural malfunction (Frangopol *et al.*, 1998).

New materials, new construction technologies and new structural systems are increasingly being used, and it is necessary to increase knowledge about their on-site performance, to control the design, to verify performance, and to create and calibrate numerical models (Bernard, 2000). Monitoring certainly provides for answers to these requests.

Monitoring can discover hidden (unknown) structural reserves and, consequently, allows for better exploitation of traditional materials and better exploitation of existing structures. In this case, the same structure can accept a higher load; that is, more performance is obtained without construction costs.

Finally, monitoring helps prevent the social, economical, ecological and aesthetical impact that may occur in the case of structural deficiency.

1.1.4 Whole Lifespan Monitoring

Monitoring should not be limited to structures with recognized deficiencies. First, because when structural deficiency is recognized, the structure functions with limited performance and the economic losses are already generated. Second, the history of events that lead to structural deficiency is not registered and it may be difficult to make a diagnosis. Third, the information concerning the health state is important as a reference, notably for complex structures where direct comparison of structural behaviour with design and numerical models does not allow for certain detection of a malfunction. That is why whole lifespan monitoring, which includes all the important phases in the structure's life, is highly recommended (Glišić *et al.*, 2002a).

Construction is a very delicate phase in the life of a structure. In particular, for concrete structures, material properties change through ageing. It is important to know whether or not the required values are achieved and maintained. Defects (e.g. premature cracking) that arise during construction may have serious consequences for structural performance (Bernard, 2000). Monitoring data help engineers to understand the real behaviour of a structure, and this leads to better estimates of real performance and, if required, more appropriate remedial action. Installation of monitoring systems during the construction phase allows monitoring to be carried out during the whole life of the structure. Since most structures have to be inspected several times during service, the best way to decrease the costs of monitoring and inspection is to install the monitoring system from the beginning.

Some structures have to be tested before service for safety reasons. At this stage, the required performance levels have to be reached. Typical examples are bridges and stadiums: the load is positioned at critical places (following the influence lines) and the parameters of interest (such as deformation, strain, displacement, rotation of section and crack opening) are measured (Hassan, 1994). Tests are performed in order to understand the real behaviour of the structure and to compare it with theoretical estimates. Monitoring during this phase can be used to calibrate numerical models that describe the behaviour of structures.

The service phase is the most important period in the life of a structure. During this phase, construction materials are subjected to degradation by ageing. Concrete cracks and creeps, and steel oxidizes and may crack due to fatigue loading. The degradation of materials is caused by mechanical (loads higher than theoretically assumed) and physico-chemical factors (corrosion of steel, penetration of salts and chlorides in concrete, freezing of concrete, etc.). As a consequence of material degradation, the capacity, durability and safety of a structure decreases. Monitoring during service provides information on structural behaviour under predicted loads, and also registers the effects of unpredicted overloading. Data obtained by monitoring is useful for damage detection, evaluation of safety and determination of the residual capacity of structures. Early damage detection is particularly important because it leads to appropriate and timely interventions. If the damage is not detected, then it continues to propagate and the structure no longer guarantees required performance levels. Late detection of damage results in either very elevated refurbishment costs (Frangopol *et al.*, 1998) or, in some cases, the structure has to be closed and dismantled. In seismic areas, the importance of monitoring is most critical.

Material degradation and/or damage are often the reasons for refurbishing existing structures. Also, new functional requirements for a structure (e.g. enlarging of bridges) lead to requirements for strengthening. For example, if strengthening elements are made of new concrete, then good interaction of the new concrete with the existing structure has to be assured: early age deformation of new concrete creates built-in stresses and bad cohesion causes delamination of the new concrete, thereby erasing the beneficial effects of the repair efforts. Since newly created structural elements that are observed separately represent new structures, the reasons for monitoring them are the same as for new structures. The determination of the success of refurbishment or strengthening is an additional justification (Inaudi *et al.*, 1999a).

When the structure no longer meets the required performance level and when the costs of reparation or strengthening are excessively high, then the ultimate lifespan of the structure is attained and the structure should be dismantled. Monitoring helps in dismantling structures safely and successfully.

1.2 The Structural Health Monitoring Process

1.2.1 Core Activities

The core activities of the structural monitoring process are: selection of monitoring strategy, installation of monitoring system, maintenance of monitoring system, data management and closing activities in the case of interruption of monitoring (Glišić and Inaudi, 2003a). Each of these activities can be split in to sub-activities, as presented in Table 1.2.

Each of the core activities is very important, but the most important is to create a good monitoring strategy. The monitoring strategy is influenced by each of the other core activities and sub-activities and consists of:

1. Establishing the monitoring aim
2. Identifying and selecting representative parameters to be monitored
3. Selecting appropriate monitoring systems
4. Designing the sensor network
5. Establishing the monitoring schedule
6. Planning data exploitation
7. Costing the monitoring.

To start a monitoring project, it is important to define the goal of the monitoring and to identify the parameters to be monitored. These parameters have to be properly selected in a way that reflects the structural behaviour. Each structure has its own particularities and, consequently, its own selection of parameters for monitoring.

There are different approaches to assessing the structure that influence the selection of parameters. We can classify them in three basic categories, namely static monitoring, dynamic monitoring, and system identification and modal analysis, and these categories can be combined. Each approach is characterized by advantages and challenges, and which one (or ones) will be used depends mainly on the structural behaviour and the goals of monitoring.

Table 1.2   Breakdown structure of the core monitoring activities

Monitoring strategy	Installation of monitoring system	Maintenance of monitoring system	Data management	Closing activities
• Monitoring aim	• Installation of sensors	• Providing for electrical supply	• Execution of measurements (reading of sensors)	• Interruption of monitoring
• Selection of monitored parameters	• Installation of accessories (connection boxes, extension cables, etc.)	• Providing for communication lines (wired or wireless)	• Storage of data (local or remote)	• Dismantling of monitoring system
• Selection of monitoring systems	• Installation of reading units	• Implementation of maintenance plans for different devices	• Providing for access to data	• Storage of monitoring components
• Design of sensor network	• Installation of software	• Repairs and replacements	• Visualization	
• Schedule of monitoring	• Interfacing with users		• Export of data	
• Data exploitation plan			• Interpretation	
• Costs			• Data analysis	
			• The use of data	

Each approach can be performed during short and long periods, permanently (continuously) or periodically. The schedule and pace of monitoring depend on how fast the monitored parameters change in time. For some applications, periodic monitoring gives satisfactory results, but information that is not registered between two inspections is lost forever. Only continuous monitoring during the whole lifespan of the structure can register its history, help to understand its real behaviour and fully exploit the monitoring benefits.

Monitoring consists of two aspects: measurement of the magnitude of the monitored parameter and recording the time and value of the measurement. In order to perform a measurement and to register it, one can use different types of apparatus. The set of all the devices destined to carry out a measurement and to register it is called a monitoring system. Nowadays, there is a large number of monitoring systems, based on different functioning principles. In general, however, they all have similar components: sensors, carriers of information, reading units, interfaces and data management subsystems (managing software). These components are presented in more detail in Chapter 2.

The Selection of a monitoring system depends on the monitoring specifications, such as the monitoring aim, selected parameters, accuracy, frequency of reading, compatibility with the environment (sensitivity to electromagnetic interference, temperature variations, humidity, ...), installation procedures for different components of the monitoring system, possibility of automatic functioning, remote connectivity, manner of data management and level at which the structure is to be monitored (i.e. global structural or local material).

For example, monitoring of new concrete structures subject to dynamic loads at the structural level can only be performed using sensors that are not influenced by local material defects or discontinuities (such as cracks, inclusions, etc.). Since short-gauge sensors are subject to local influences, a good choice is to use a monitoring system based on long-gauge or distributed sensors. In addition, the sensors are to be embeddable in the concrete, insensitive to environmental conditions and the reading unit must be able to perform both static and dynamic measurements with a certain frequency and a certain accuracy.

Several parameters are often required to be monitored, such as average strains and curvatures in beams, slabs and shells, average shear strain, deformed shape and displacement, crack occurrence and quantification, as well as indirect damage detection. The use of separate monitoring systems and separate sensors for each parameter mentioned would be costly and complex from the point of view of installation and data assessment. This is why it is preferrable to use only a limited number of monitoring systems and types of sensor.

In order to extract maximum data from the system it is necessary to place the sensors in representative positions on the structure. The sensor network to be used for monitoring depends on the geometry and the type of structure to be monitored, parameters and monitoring aims. The design of sensor networks is developed and presented in Chapters 4 and 5.

The installation of the monitoring system is a particularly delicate phase. Therefore, it must be planned in detail, seriously considering on-site conditions and notably the structural component assembly activities, sequences and schedules.

The components of the monitoring system can be embedded (e.g. into the fresh concrete or between the composite laminates), or installed on the structure's surface using fastenings, clamps or gluing. The installation may be time consuming, and it may delay construction work if it is to be performed during construction of the structure. For example, components of a monitoring system that are to be installed by embedding in fresh concrete can only be safely installed during a short period between the rebar completion and pouring of concrete. Hence, the installation schedule of the monitoring system has to be carefully planned to take into account the schedule of construction works and the time necessary for the system installation. At the same time, one has to be flexible in order to adapt to work schedule changes, which are frequent on building sites.

When installed, the monitoring system has to be protected, notably if monitoring is performed during construction of the structure. Any protection has to prevent accidental damage during the construction and ensure the longevity of the system. Thus, all external influences, periodic or permanent, have to be taken into account when designing protection for the monitoring system.

Structures have different life periods: construction, testing, service, repair and refurbishment, and so on. During each of these periods, monitoring can be performed with an appropriate schedule of measurements. The schedule of measurements depends on the expected rate of change of the monitoring parameters, but it also depends on safety issues. Structures that may collapse shortly after a malfunction occurs must be monitored continuously, with maximum frequency of measurements. However, the common structures are designed in such a manner that collapse occurs only after a significant malfunction that develops over a long period. Therefore, in order to decrease the cost of monitoring, the measurements can be preformed less frequently, depending on the expected structural behaviour. An example is given below for static monitoring of concrete structures:

- *Early and very early age of concrete.* Possible only if low-stiffness sensors are embedded in the concrete (Glišić, 2000). The monitoring schedule of early-age deformation is one to four sessions of measurements per hour during the first 24–36 h and four measurements per day to one measurement per week afterwards, depending on concrete evolution ('session' means one measurement for each sensor).
- *Continuous monitoring for 24–48 h.* This is recommended in order to record the behaviour of the structure due to daily temperature and load variations. This session of measurements is to be performed at a pace of one measurements session per hour during 24–48 h, at least once per season of each year.
- *Construction period.* The schedule must be adapted to construction work. It is recommended to perform at least one measurement session after each construction step that changes the loads in previously built elements (pouring of new storeys of a building, assembling of elements by prestressing, transportation, etc.).
- *Testing load (if any).* Generally a minimum of one measurement session after each load step.
- *Period before refurbishment, repair or enlargement.* These measurements will serve to learn about the structural behaviour before reconstruction. They are to be performed several times per day (e.g. one session in the morning, noon, afternoon and night) during an established (representative) period. In addition, several continuous 24 h or 48 h monitoring periods (session each hour) are recommended in order to determine the daily influence of temperature and loads.
- *During refurbishment, repair or enlargement.* In general, the same schedule as for construction, combined with four times per day and 24 or 48 h sessions.
- *Long-term monitoring during service.* At least one to four sessions per day are recommended for permanent static monitoring and at least one per week to one per month for periodic static monitoring. Yearly periodic 24–48 h continuous sessions (at least one session every hour during 24 h) are also recommended.
- *Special events.* Measurement sessions during and after strong winds, heavy rain, earthquakes or terrorist acts.

The data management can be basic or advanced. Basic data management consists of execution of measurements (reading of sensors), storage of data (local or remote) and providing for access to data. The monitoring data can be collected manually, semi-automatically or automatically, on site or remotely, periodically or continuously, statically and dynamically. These options can be combined in different ways; for example, during testing of a bridge it is necessary to perform measurements semi-automatically, on site and periodically (after each load step). For long-term in-use monitoring, the maximal performance is automatic, remote (from the office), continuous collecting of data, without human intervention. Possible methods of data collection (reading of sensors) are presented schematically in Figure 1.2.

Data can be stored, for example, in the form of reports, tables and diagrams on different types of support, such as electronic files (on hard disc, CD, etc.) or hard versions (printed on paper). The manner of storage of data has to ensure that data will not be lost (data stored in a 'central library' with backups) and that prompt access to any selected data is possible (e.g. one can be interested to access only data from one group of sensors and during a selected period of monitoring). The possible manners of storage and access to data are presented in Figure 1.3.

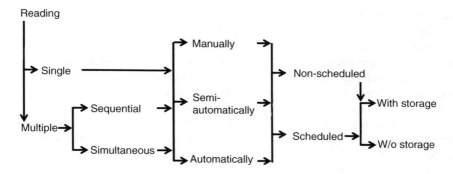

Figure 1.2 Methods of collecting the data (courtesy of SMARTEC).

The software that manages the collection and storage of data is to be a part of the monitoring system. Otherwise, data management can be difficult, demanding and expensive.

Advanced data management consists of interpretation, visualization, export, analysis and the use of data (e.g. generation of warnings and alarms). Collected data are, in fact, a huge amount of numbers (dates and magnitudes of monitoring parameters) and have to be transformed to useful information concerning the structural behaviour. This transformation depends on the monitoring strategy and algorithms that are used to interpret and analyse the data. This can be performed manually, semi-automatically or automatically.

Manual data management consists of manual interpretation, visualization, export and analysis of data. This is practical in cases where the amount of data is limited. Semi-automatic data management consists of a combination of manual and automatic actions. Typically, export of data is manual and analysis is automatic, using an appropriate software. This is applicable in cases where the data analysis is to be performed only periodically. Automatic data management is the most convenient, since it can be performed rapidly and independent of data amount or frequency of analysis. Finally, based on information obtained from data analysis, planned actions can be undertaken (e.g. warnings can be generated and exploitation of the structure stopped in order to guarantee safety).

The data management has to be planned along with the selection of the monitoring strategy. Appropriate algorithms and tools compatible with the chosen monitoring system have to be selected.

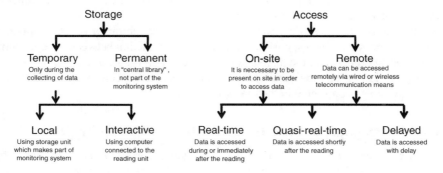

Figure 1.3 Possible methods of storage and access to data (courtesy of SMARTEC).

The monitoring strategy is often limited by the budget available. From a monitoring performance point of view, the best is to use powerful monitoring systems, dense sensor networks (many sensors installed in each part of the structure), software allowing remote and automatic operation. On the other hand, the cost of such monitoring can be very elevated and unaffordable. That is why it is important to develop an optimal monitoring strategy, providing good evaluation of structural behaviour, but also affordable in terms of costs. There are no two identical structures; consequently, the monitoring strategy is different for each structure. Methods used to develop a monitoring strategy that is optimal in terms of monitoring performance and budget are presented in the following chapters of this book. Based on our experience of applying the proposed methods, an estimated budget for monitoring of a new structure ranges between 0.5 % and 1.5 % of the total cost of the structure.

1.2.2 Actors

The main actors (entities) involved in monitoring are the monitoring authority, the consultant, the monitoring companies and the contractors. These entities must collaborate closely with each other in order to create and implement an efficient and performing monitoring strategy. These entities need not necessarily to be different; for example, a monitoring company can also have a role of consultant or contractor.

The monitoring authority is the entity that is interested in and decides to implement monitoring. It is usually the owner of the structure or the entity that is, for some reason, interested in the safety of the structure (e.g. legal authority). The monitoring authority finances the monitoring and benefits from it. It is responsible for defining the monitoring aims and for approving the proposed monitoring strategy. The same authority is later responsible for maintenance and data management (directly or by subcontracting to the monitoring company or contractor).

The consultant proposes a monitoring strategy to the monitoring authority. This strategy consists of performing the necessary analysis of the structural system, estimating loads, performing numerical modelling, evaluating risks and creating another monitoring strategy if the initial one is rejected by the monitoring authority. After the delivery of the monitoring system, the consultant may perform supervision of the installation and commissioning of the monitoring system.

The company devoted to monitoring (monitoring company) is basically responsible for delivery of the monitoring system. However, the same company can often have a role of consultant (development of the monitoring strategy in collaboration with the responsible authority) or contractor (implementation of the monitoring system).

The installation of the monitoring system is performed by a contractor with the support of the monitoring company and the responsible authority. The interaction between the core activities of the monitoring process and the main actors is presented in Figure 1.4.

As an illustration of the topics and processes presented in Sections 1.1 and 1.2, an on-site monitoring example is presented in the next section.

1.3 On-Site Example of Structural Health Monitoring Project

Once every generation, Switzerland treats itself to a national exhibition commissioned by the Swiss Confederation. Expo 02 was spread out over five temporary *arteplages* built on and around Lake Biel, Lake Murten and Lake Neuchâtel, located in the northwest of Switzerland

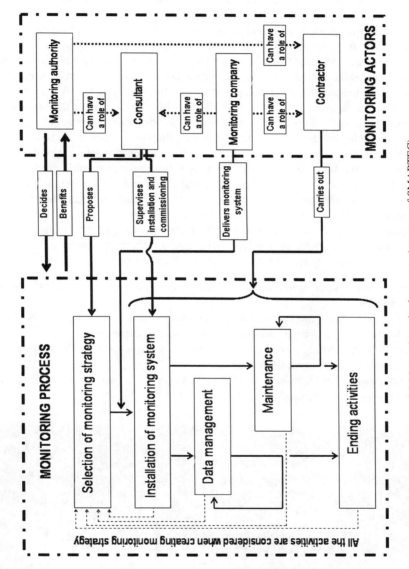

Figure 1.4 Interaction between monitoring core activities and monitoring actors (courtesy of SMARTEC).

11

(Cerulli *et al.*, 2003). Each *arteplage* was related to a particular theme, which was reflected in its architecture and exhibitions. The 'arteplage' at Neuchâtel was related to 'Nature and Artificiality'; a big steel and wooden whale eating a village represented *The Adventures of Pinocchio* fairy tale from the Italian writer Collodi. The belly of the whale held an exposition dedicated to robotic and artificial intelligence, while the rest of the village was developed on two floors with steel piles/beams and wooden walls and floors. The 'Piazza Pinocchio' was built together with other exposition buildings on one large artificial peninsula (platform), approximately 50 m from the shore and 5 m above the lake water level. A large textile membrane was used to cover the Piazza Pinocchio. After Expo 02, the peninsula was dismantled. The global views of Expo 02 in Neuchâtel and the whale structure are shown in Figure 1.5.

The peninsula consisted of a steel grid platform structurally supported by underwater steel columns. One of the architects' aims was to allow visitors to walk over the two exposition floors without restrictions. A concentration of visitors at one exhibition place, combined with temperature variations and differential settlements of columns, could create a redistribution in the structural elements that would be difficult to predict. Numeric simulation of the structural behaviour would have been too laborious without giving an indisputable feedback on the real structural behaviour. In order to ensure structural safety and optimal serviceability of the peninsula structure during the opening and in service, the Expo 02 committee (monitoring authority) decided to monitor the Piazza Pinocchio.

The monitoring company selected also had the roles of consultant and contractor; that is, the company was also in charge of developing the monitoring strategy and implementing the monitoring system. The monitoring specifications were as follows:

1. To ensure structural safety and optimal serviceability of the peninsula structure during the opening and in service
2. Identified representative monitored parameters are normal (axial) forces in the columns; they are determined for average strain and temperature monitoring
3. An optical-fibre monitoring system with high accuracy allowing for quasi-real-time, automatic and remote operation was selected
4. A so-called scattered simple topology combined with parallel topologies was used to monitor the columns (see Figure 1.6 and Sections 4.2 and 4.3)

Figure 1.5 View of the artificial peninsula hosting the 'Piazza Pinoccio' (left) and whale structure (right) (courtesy of SMARTEC).

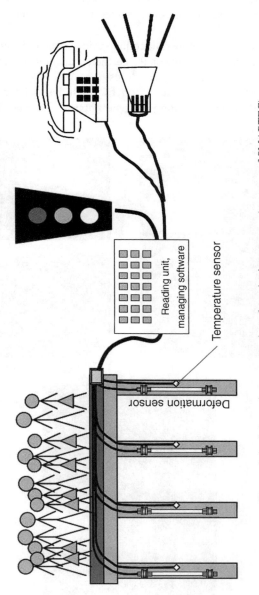

Figure 1.6 Schematic representation of monitoring strategy (courtesy of SMARTEC).

5. Monitoring is performed continuously during the exposition's opening hours
6. The data received from the monitoring is used to stop overloading of the platform by visitors and to evacuate the exhibition area in the case of structural malfunction
7. To make monitoring costs affordable, taking into account the temporary purpose of the structure, the monitoring system was simply rented from the monitoring company.

The monitoring strategy was developed in collaboration with engineers responsible for the structural design, with architects to decide on the aesthetics and logistics, and with the Expo 02 Security Department to develop warning procedures.

The technical aims were to enable detection of small load changes, to identify thermally induced strains and to detect bending on representative columns. The resolution of the

Sensor fastening detail

Sensor fastened to the column web   Temp.sensor fastening detail

Sensor in the cross-section edges

Figure 1.7   Photographs taken during the installation (courtesy of SMARTEC).

monitoring system selected is 2 $\mu\varepsilon$, which allowed the detection of the weight caused by 10 people (~700 kg) carried by one column (which corresponds to about 20 kg m^{-2}). Deformation sensors with a 1 m long gauge-length were selected. To detect biaxial bending moment effects, four sensors were installed at the edges of the cross-section of one representative column. To determine thermal strain and separate it from elastic strain, compatible conventional temperature sensors were used. The monitoring concept is represented schematically in Figure 1.6.

Continuous measurements were carried out over 5 months during the daily opening hours (about 18 h per day). In the morning, before visitors were on site, a measurement was taken. This measurement was useful for comparing the measurements without live loads. After each measurement session was completed, the forces in the columns were calculated in quasi-real time and compared with predefined thresholds obtained using the algorithms developed. If the warning threshold was reached, then the alert status was activated.

People "free movement" test

People standing on representative locations simulating
concentrated load

Figure 1.8 Photographs taken during the tests (courtesy of SMARTEC).

Sensor installation was carried out in different stages. To help the main contractor to maintain the construction work schedule, the sensors were installed on columns during construction and the connecting cables were installed at a later date inside the first-floor wooden pavement.

The central measurement point consisted of one reading unit, one optical channel switch and one computer connected to the telephone line. The central measurement point was installed in the control room (on the first floor) together with other devices used to manage and control the Piazza Pinocchio's shows and performances. Photographs of the installation are presented in Figure 1.7.

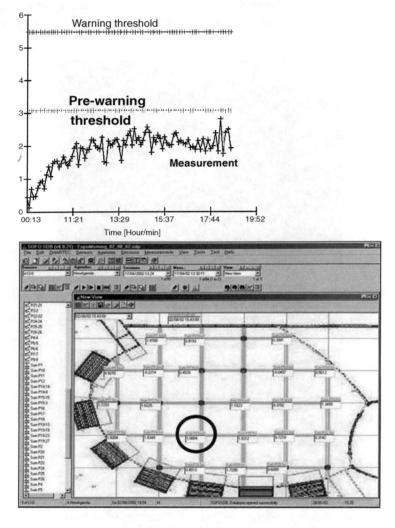

Figure 1.9   Visualization of a single measurement (left) and plan view of whale floor with 'windows' showing the actual value of the force in the corresponding column; if the threshold is reached, the colour of the window changes to yellow (pre-warning) or red (warning) (courtesy of SMARTEC).

Since some sensors were installed in rooms accessible to visitors, it was necessary to hide them in order to provide good aesthetical impact and protection. Moreover, neon lamps were installed in certain columns, so protection against unintentional accident was necessary. For these reasons the architects decided to protect the column by using an aluminium grating. The thermocouple heads were covered using polystyrene to provide ambient thermal isolation.

Before the national exposition started, the committee decided to test the structure and the monitoring system. More than 1000 people had been asked to visit the exposition area freely and to consent to a trial load test, where people had to stand very closely for a few minutes at certain locations. The tests were performed with high safety precautions. The monitoring system passed the tests successfully and was commissioned and put in service. Photographs taken during the tests are shown in Figure 1.8.

The data management consists of sensor readings, analysis of results, storage of results on a local computer, comparison with predefined thresholds and visualization of both measured values and warnings. To enable access to the monitoring system from different locations, the remote monitoring option was provided via a telephone line. Every day, at closing time, the system automatically executed a backup of the database and generated an Excel file (as an official results document). After that, it prepared the new configuration file to be used the following morning and switched off. Examples of data visualizations are given in Figure 1.9.

After Expo 02 closed the peninsula structure and the monitoring system were dismantled. The monitoring system was returned to the monitoring company.

An example of the complete monitoring process and interaction with monitoring actors has been presented in order to illustrate the notions developed and presented in the previous two sections.

2

Fibre-Optic Sensors

2.1 Introduction to Fibre-Optic Technology

A typical health-monitoring system is composed of a network of sensors that measure the parameters relevant to the state of the structure and its environment. For civil structures, such as bridges, tunnels, dams, geostructures, power plants, high-rise buildings and historical monuments, the most relevant parameters are:

- Position, deformations, inclinations, strains, forces, pressures, accelerations and vibrations
- Temperatures
- Humidity, pH and chlorine concentration
- Environmental parameters, such as air temperature, wind speed and direction, solar irradiation, precipitation, snow accumulation, water levels and flow, pollutant concentration.

Conventional sensors based on mechanical and/or electrical transducers are able to measure most of these parameters. In the last few years, fibre-optic sensors have made a slow but significant entrance in the sensor panorama. After an initial euphoric phase when optical-fibre sensors seemed on the verge of invading the whole world of sensing, it now appears that this technology is manly attractive in those cases where it offers superior performance compared with the more proven conventional sensors. The additional value can include improved quality of the measurements, better reliability, the possibility of replacing manual readings and operator judgment with automatic measurements, an easier installation and maintenance or a lower lifetime cost.

Even though fibre-optic sensors are apparently expensive for widespread use in health monitoring, they are, however, better approaches for applications where reliability in challenging environments is essential. It is sometimes crucial to use a reliable technology for critical health monitoring: price is often no longer a showstopper when the security or efficient management of very expensive systems, such as civil engineering structures, could lead to catastrophic consequences.

In commercial applications where fibre-optic sensors have been selected for health monitoring, the benefit often more than justifies the higher cost of this type of technology. Fibre-optic sensors can even become cost effective when involving a significant number of sensors, such as

Fibre Optic Methods for Structural Health Monitoring   B. Glišić and D. Inaudi
© 2007 John Wiley & Sons, Ltd

in civil engineering applications, using quasi-distributed or fully distributed fibre-optic sensors. In some extreme applications, such as in the oil and gas industry, fibre-optic sensors are sometimes the only available solution for reliable and long-term physical parameter monitoring.

The greatest advantages of the fibre-optic sensors are intrinsically linked to the optical fibre, which is either simply a link between the sensor and the signal conditioner, or is the sensor itself in long-gauge and distributed sensors. In almost all fibre-optic sensor applications, the optical fibre is a thin glass fibre that is usually protected mechanically with a polymer coating (or even a metal coating in extreme cases) and often inserted in a cable designed to be suitable for targeted applications. Glass, since it is an inert material very resistant to almost all chemicals even at elevated temperatures, is an ideal material for applications in harsh chemical environments, such as those encountered in oil and gas wells, sparkplug engines or a concrete structure. It is also interesting since it is resistant to weathering effects and it is not subject to any corrosion. The latter property is a great advantage for long-term reliable health monitoring of civil engineering structures. Some identified problems, such as optical-fibre hydrogen darkening in oil wells, are not necessarily an issue when selecting the appropriate interrogation technology and fibre type.

Since the light confined in the core of the optical fibres used for sensing purposes does not interact with any surrounding electromagnetic (EM) field, fibre-optic sensors are therefore intrinsically immune to any EM interference (EMI). With such a unique advantage over their electrical counterparts, fibre-optic sensors are obviously the ideal sensing solution when the presence of EM, radio frequency (RF) or microwaves (MW) cannot be avoided. For instance, fibre-optic sensors will not be affected by any EM field generated by lightning hitting a monitored bridge or dam, unless the fibre is damaged thermally. Optical-fibre sensors are also not affected by nearby electrical machinery, such as electric locomotives, power lines or transformers. Besides increasing sensor reliability, its EMI immunity could, for instance, be a unique advantage for monitoring hot spots in high-power electrical transformers monitored with fibre-optic temperature sensors. By design, fibre-optic sensors are intrinsically safe and naturally explosion proof, making them particularly suitable for health monitoring applications of risky civil structures, such as gas pipelines or chemical plants.

Probably the greatest advantage of fibre-optic sensors is still their small size. In most cases, the diameter of bare fibre-optic sensors, usually in the range 125–500 μm, is very appropriate in space-restricted environments, such as thin composite structures. Most frequently, a fibre-optic sensor has an axial geometry suitable also for many applications where this is a benefit, such as instrumentation of bolts or similar cylindrical devices.

The ability to measure over distances of several tens of kilometres without the need for any electrically active component is also an advantage inherited from the fibre-optic telecommunications industry. This is an important feature when monitoring large and remote structures, such as pipelines or multiple bridges along a single highway.

Fibre-optic sensors offer a great variety of parameters that can be measured, so that multiple parameters can be mixed on the same network.

Compared with conventional electrical sensors, fibre-optic sensors offer new and unique sensing topologies, including in-line multiplexing and fully distributed sensing, offering novel monitoring opportunities.

Finally, the tremendous developments in the optical telecommunications market have reduced considerably the cost and increased the performances of optical fibres and their associated optical components. However, the fibre-optic sensor segment, which is still too small an industry to justify by itself all the investments made so far in the telecoms field, takes advantage of the progress made in optical communications.

The fibre-optic sensor market is growing significantly and, as predicted by several market studies, will continue to do so in the future. Certainly, high-volume applications for fibre-optic sensors have to be addressed commercially in order to be more and more competitive with the usually less expensive and very familiar solutions using electrical sensors. In particular, demanding applications where conventional sensors are difficult to apply present the best opportunities for fibre-optic sensors. Among such applications, those involving health monitoring of civil engineering structures probably offer the best opportunities for the different technologies based on fibre-optic sensors.

The first successful industrial applications of fibre-optic sensors to civil structural monitoring demonstrate that this technology is now sufficiently mature for routine use and that it can compete as a peer with conventional instrumentation.

A typical health monitoring system is composed of several elements that are equally important to achieve an effective system.

1. *Sensors*. These transducers convert the parameter to be measured to a different and measurable quantity. In the case of optical-fibre sensors, the sensing element typically transforms a change in the monitoring parameter in a corresponding change in the properties of the light guided by the optical fibre. Such a change can involve its intensity, phase, spectral content, polarization state or a combination of these.
2. *Cable network*. This is used to connect the sensors to the data acquisition (DAQ) system. Fibre-optic sensors offer the advantage of a purely passive cable network that is composed entirely of optical fibres. Several signals from multiple sensors can sometimes be combined into a single optical fibre or into a multi-fibre cable.
3. *DAQ system*. For each sensor type, we find a corresponding DAQ unit that observes the change in the optical signal into intelligible information about the original change in the structure. The data are typically made available in digital format and already incorporate additional information about the calibration curve of the sensor.
4. *Data management system*. Data must be stored in an organized way, so that it can be properly analysed later. Nowadays, the best way to store data for long-term monitoring purposes is in a relational database. However, it is important to ensure that the data are properly stored and duplicated to avoid accidental loss of data.
5. *Data analysis*. This layer analyses the data to transform it into information that can be used for decision-making purposes. This element can also include tools for data presentation and publishing.

2.2 Fibre-Optic Sensing Technologies

There is a large variety of fibre-optic sensors for health monitoring, developed by both academic and industrial institutions. Universities and industrial research centres are developing and producing a large variety of sensors for the most diverse types of measurements and applications. In this overview, we will concentrate on sensors for health monitoring that have reached an industrial level or are at least at the stage of advanced field trials. This chapter is not intended to give a detailed description of each technology, but rather to serve as a general overview

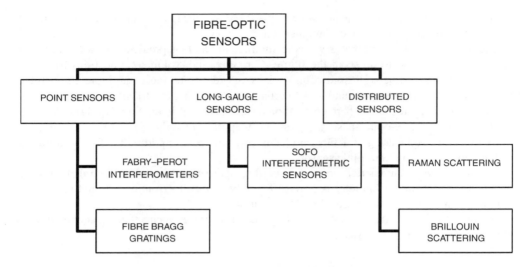

Figure 2.1 Classification of fibre-optic sensing technologies.

of the main technologies and implementation issues. The interested reader should refer to the literature cited and to the documentation of the manufacturers of the different systems to get additional details.

Figure 2.1 classifies the fibre-optic sensing technologies according to the measurement principle and Table 2.1 summarizes the main characteristics of these technologies, which are discussed in more detail in the following sections.

2.2.1 SOFO Interferometric Sensors

The SOFO system (SOFO is an acronym derived from the French for 'structural monitoring by optical fibres', i.e. *surveillance d'ouvrages par fibres optiques*) is a long-gauge fibre-optic deformation sensor with a resolution in the micrometre range, an excellent long-term stability and insensitivity to temperature. It was developed at the Swiss Federal Institute of Technology in Lausanne (EPFL) and is now commercialized by SMARTEC and the Roctest Group. The functional principles of the SOFO system are schematized in Figure 2.2.

The sensor consists of a pair of single-mode fibres installed in the structure to be monitored. One of the fibres, called the measurement fibre, is in mechanical contact with the host structure itself (being attached to it at its two extremities and prestressed in-between) and the other, the reference fibre, is placed loose in the same pipe. All deformations of the structure will then result in a change of the length difference between these two fibres.

To make an absolute measurement of this path imbalance, a low-coherence double Michelson interferometer is used. The first interferometer is made of the measurement and reference fibres, and the second is contained in the portable reading unit. This second interferometer, by means of a scanning mirror, can introduce a well-known path imbalance between its two arms.

Because of the reduced coherence of the source used (the 1.3 μm radiation of a light-emitting diode), interference fringes are detectable only when the reading interferometer compensates exactly the length difference between the fibres in the structure.

Table 2.1 Summary of fibre-optic sensing types and typical performances

	SOFO interferometric	Fabry–Perot interferometric	Fibre Bragg gratings	Raman scattering	Brillouin scattering
Sensor type	Long-gauge (integral strain)	Point	Point	Distributed	Distributed
Main measurable parameters	Deformation Strain Tilt Force	Strain Temperature Pressure	Strain Temperature Acceleration Water level	Temperature	Strain Temperature
Multiplexing	Parallel	Parallel	In-line and parallel	Distributed	Distributed
Measurement points in one line	1	1	10–50	10 000	30 000
Typical accuracy — Strain ($\mu\varepsilon$)	1	1	1		20
Typical accuracy — Deformation (μm)	1	100	1		
Typical accuracy — Temperature ($^{\circ}$C)		0.1	0.1	0.1	0.2
Typical accuracy — Tilt (μrad)	30				
Typical accuracy — Pressure (% full scale)		0.25			
Range	20 m gauge			8 km	30 km, 150 km with range extenders
Fibre type	Single mode	Multimode	Single mode	Multimode	Single mode

If this measurement is repeated at successive times, the evolution of the deformations in the structure can be followed without the need for continuous monitoring. This means that a single reading unit can be used to monitor several fibre pairs in multiple structures.

The precision and stability obtained by this setup have been quantified in laboratory and field tests to 2 μm, independently of the sensor length over more than 10 years. Even a change in the

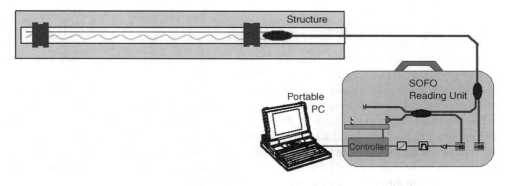

Figure 2.2 Setup of the SOFO interferometric sensor system (courtesy of SMARTEC).

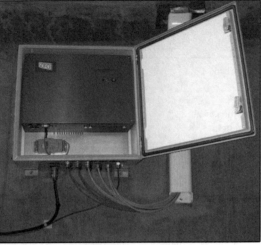

Figure 2.3   SOFO reading unit, portable and for permanent installation (courtesy of SMARTEC).

fibre transmission properties does not affect the precision, since the displacement information is encoded in the coherence of the light and not in its intensity.

The reading unit is portable, waterproof and battery powered, making it ideal for dusty and humid environments, such as those found on most building sites (see Figure 2.3). Each measurement takes about 10 s and all the results are automatically analysed and stored for further interpretation by an external laptop computer.

The measurements can either be performed manually, by connecting the different sensors one after the other, or automatically by means of an optical switch. Since the measurement of the length difference between the fibres is absolute, there is no need to maintain a permanent connection between the reading unit and the sensors. A single unit, therefore, can be used to monitor multiple sensors and structures with the desired frequency.

The SOFO system has been used successfully to monitor more than 400 structures so far, including bridges, buildings, tunnels, piles, anchored walls, dams, historical monuments, nuclear power plants and laboratory models, proving to be one of the most adapted and widely used fibre-optic technologies for civil structural monitoring.

2.2.2 Fabry–Perot Interferometric Sensors

Extrinsic Fabry–Perot interferometers (EFPIs) are constituted by a capillary silica tube containing two cleaved optical fibres facing each other (see Figure 2.4), but leaving an air gap of a few micrometres or tens of micrometres between them (Measures, 2001). When light is launched into one of the fibres, a back-reflected interference signal is obtained. This is due to the reflection of the incoming light on the glass-to-air and on air-to-glass interfaces. This interference can be demodulated using coherent or low-coherence techniques to reconstruct the changes in the fibre spacing. Since the two fibres are attached to the capillary tube near its two extremities (with a typical spacing of 10 mm), the gap change will correspond to the average strain variation between the two attachment points.

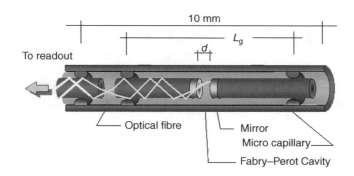

Figure 2.4   Functional principles of Fabry–Perot sensors (courtesy of FISO).

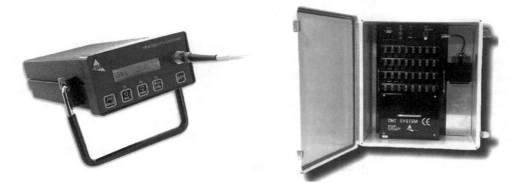

Figure 2.5   Demodulators for Fabry–Perot interferometers for single and multiple channels (courtesy of Roctest) .

Figure 2.5 shows some typical demodulators for Fabry–Perot sensors.

2.2.3 Fibre Bragg-Grating Sensors

Bragg gratings are periodic alterations in the index of refraction of the fibre core that can be produced by adequately exposing the fibre to intense UV light. The gratings produced typically have lengths of the order of 10 mm. If a tuneable light source is injected in to the fibre containing the grating, then the wavelength corresponding to the grating pitch will be reflected while all other wavelengths will pass through the grating undisturbed, as depicted in Figure 2.6. Since the grating period is strain and temperature dependent, it becomes possible to measure these two parameters by analysing the intensity of the reflected light as a function of the wavelength. This is typically done using a tuneable laser containing a wavelength filter (such as a Fabry–Perot cavity) or a spectrometer.

Resolutions of the order of 1 $\mu\varepsilon$ and 0.1 °C can be achieved with the best demodulators. If strain and temperature variations are expected simultaneously, then it is necessary to use a free reference grating that measures the temperature alone and use its reading to correct the

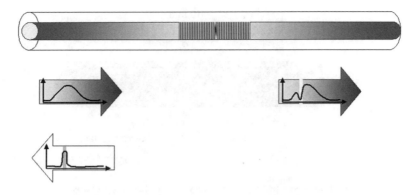

Figure 2.6   Functional principle of fibre Bragg-grating (FBG) sensors (courtesy of SMARTEC).

strain values. Setups allowing the simultaneous measurement of strain and temperature have
been proposed, but have yet to prove their reliability in field conditions. The main interest in
using Bragg gratings resides in their multiplexing potential. Many gratings can be written in
the same fibre at different locations and tuned to reflect at different wavelengths. This allows
the measurement of strain at different places along a fibre using a single cable, as shown in
Figure 2.7. Typically, 4–16 gratings can be measured on a single fibre line. It has to be noted
that, since the gratings have to share the spectrum of the source used to illuminate them, there
is a trade-off between the number of gratings and the dynamic range of the measurements on
each of them.

A large number of measurement techniques and instruments for FBG demodulation are
available, differing in terms of wavelength accuracy and range as well as in the dynamic
sensing properties (Ferdinand *et al.*, 1994, 1997; Kersey, 16). Commercial demodulators are
available from MicronOptics (USA), SMARTEC (Switzerland) and many other companies
worldwide.

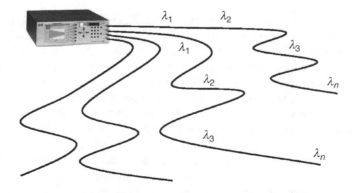

Figure 2.7   Parallel and in-line multiplexing of FBG sensors (courtesy of SMARTEC).

2.2.4 Distributed Brillouin- and Raman-Scattering Sensors

Developed for telecommunication applications, optical time-domain reflectometers (OTDRs) have been the starting point of distributed sensing techniques. They use the Rayleigh-scattered light to measure the attenuation profiles of long-haul fibre-optics links. In the OTDR technique, an optical pulse is launched in the fibre and a photodetector measures the amount of light that is backscattered as the pulse propagates down the fibre. The detected signal, the so-called Rayleigh signature, presents an exponential decay with time that is directly related to the linear attenuation of the fibre. The time information is converted to distance information if the speed of light is known, similar to radar detection techniques. In addition to the information on fibre losses, the OTDR profiles are very useful to localize breaks, to evaluate splices and connectors and, in general, to assess the overall quality of a fibre link.

Raman- and Brillouin-scattering phenomena have been used for distributed sensing applications over the past few years. Raman scattering was first proposed for sensing applications in the 1980s, whereas Brillouin scattering was introduced later as a way to enhanced the range of OT-DRs and then for strain and/or temperature monitoring applications. Figure 2.8 schematically shows the spectrum of the scattered light from a single wavelength λ_0 in optical fibres. Both Raman- and Brillouin-scattering effects are associated with different dynamic inhomogeneities in the silica and, therefore, have completely different spectral characteristics.

The Raman-scattered light is caused by thermally influenced molecular vibrations. Consequently, the backscattered light carries the information on the local temperature where the scattering occurred. The amplitude of the anti-Stokes component is strongly temperature dependent, whereas the amplitude of the Stokes component is not.

The Raman sensing technique requires some filtering to isolate the relevant frequency components. This consists of recording the ratio between the anti-Stokes amplitude and the Stokes amplitude, which contains the temperature information. Since the magnitude of the spontaneous Raman backscattered light is quite low (10 dB below spontaneous Brillouin scattering), high numerical aperture multimode fibres are used in order to maximize the guided intensity of the backscattered light. However, the relatively high attenuation characteristics of multimode fibres limit the distance range of Raman-based systems to approximately 8 km.

Brillouin scattering (Karashima *et al.*, 1990) occurs because of an interaction between the propagating optical signal and thermally excited acoustic waves in the gigahertz range present

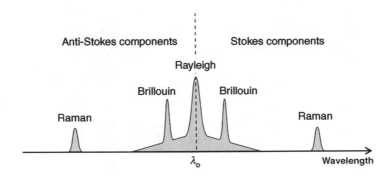

Figure 2.8 Optical scattering components in optical fibres.

in the silica fibre, giving rise to frequency-shifted components. It can be seen as the diffraction of light on a dynamic grating generated by an acoustic wave (an acoustic wave is actually a pressure wave that introduces a modulation of the index of refraction through the elasto-optic effect). The diffracted light experiences a Doppler shift, since the grating propagates at the acoustic velocity in the fibre. The acoustic velocity is directly related to the density of the medium that is temperature and strain dependent. As a result, the so-called Brillouin frequency shift carries the information about the local temperature and strain of the fibre.

The active stimulation of Brillouin scattering can be achieved by using two optical light waves. In addition to the optical pulse, usually called the pump, a continuous wave (CW) optical signal, the so-called probe signal, is used to probe the Brillouin frequency profile of the fibre (Niklès *et al.*, 1994, 1997). A stimulation of the Brillouin scattering process occurs when the frequency difference (or wavelength separation) of the pulse and the CW signal corresponds to the Brillouin shift (resonance condition) and provided that both optical signals are counter-propagating in the fibre. The interaction leads to a larger scattering efficiency, resulting in an energy transfer from the pulse to the probe signal and an amplification of the probe signal. The frequency difference between the pulse and probe can be scanned for precise and global mapping of the Brillouin shift along the sensing fibre. At every location, the maximum of the Brillouin gain is computed and the information transformed to temperature or strain using the appropriate calibration coefficients. The probe signal intensity can be adjusted to acceptable levels for low-noise fast acquisition whatever the measurement conditions and fibre layout, thus solving the main problem that is generally associated with distributed sensing based on spontaneous light scattering.

Measurement systems based on Brillouin scattering are commercially available from different vendors, including Omnisens (Switzerland), SMARTEC (Switzerland), Sensornet (UK), ANDO (Japan) and Nubrex (Japan). A Brillouin interrogator is shown in Figure 2.9.

For applications where the sensing area is located more than 25 km away from the instrument, it is possible to boost the optical signal using light amplifiers: the so-called range extender modules. The module performs active signal regeneration by using optical amplification techniques similar to those extensively used in optical telecommunications. The modules can be

Figure 2.9   DiTeSt Brillouin interrogator (courtesy of Omnisens).

cascaded, leading to remote distances in excess of hundreds of kilometres. These modules are far simpler than the interrogator system itself and, therefore, allow the monitoring of longer distances with reduced cost. Furthermore, the range extender modules themselves are more amenable to operation in challenging environments (e.g. subsea), requiring lower power and being based on a few reliable components.

For temperature measurements, the Brillouin sensor is a strong competitor to systems based on Raman scattering, whereas for strain measurements it has practically no rivals.

Since the Brillouin frequency shift depends on both the local strain and temperature of the fibre, the sensor setup will determine the actual sensitivity of the system. For measuring temperatures, it is sufficient to use a standard telecommunications cable. These cables are designed to shield the optical fibres from an elongation of the cable. The fibre, therefore, will remain in its unstrained state and the frequency shifts can be unambiguously assigned to temperature variations. If the frequency shift of the fibre is known at a reference temperature, then it will be possible to calculate the absolute temperature at any point along the fibre. Measuring distributed strains requires a specially designed sensor. A mechanical coupling between the sensor and the host structure along the whole length of the fibre has to be guaranteed. To resolve the cross-sensitivity to temperature variations, it is also necessary to install a reference fibre along the strain sensor. As with the temperature case, knowing the frequency shift of the unstrained fibre allows an absolute strain measurement. Distributed-sensing cable designs are discussed later in this chapter.

Using Raman scattering it is possible to obtain distributed temperature measurements over a maximum length of typically 8 km. Contrary to Brillouin scattering, no strain measurement is possible. The Raman scattering produces two broadband components at higher and lower frequencies than the exciting pump wave. Measuring the intensity ratio between these bands, called the Stokes and anti-Stokes emissions, it is possible to calculate the temperature at any given point along the fibre line. Typical spatial resolutions are of the order of 1 m and temperature resolution is 0.2 °C.

Raman systems are available commercially from several companies, including SMARTEC, Agilent, and Sensornet, and have been used in different structural monitoring applications, in particular for oil wells and for the monitoring of leakage in earth dams (Hurtig *et al.*, 1996). A Raman interrogator is shown in Figure 2.10.

Figure 2.10 Distributed Raman-scattering interrogator (courtesy of Sensornet).

2.3 Sensor Packaging

Owing to the variety of materials and environmental conditions that can be found in the different fields where fibre-optic sensors can be applied, it is difficult to give general guidelines for the installation of the sensors and the optical links. However, some critical points are common to the different fields and should be considered attentively in the design or selection process. The first concern is the installation of the sensors themselves into or onto the host material (Brönnimann *et al.*, 1998). On the one hand, it is necessary to guarantee a good mechanical contact between the fibre sensor and the structure, while on the other hand it is important to protect the fibres mechanically (Inaudi *et al.*, 1994). In the case of short-gauge strain sensors it is difficult to add additional layers of protection to the fibres without altering the sensor response. In this case, the sensor has to be adhered to or embedded in the structure directly. For example, it is possible to glue a Bragg grating sensor to the rebars before concrete is poured or to embed the sensor into a composite material. In other cases the strain sensor can be first embedded in a buffer material that is mechanically compatible with the surrounding material (e.g. a mortar prism for the installation in a concrete material) (Habel and Hofmann, 1994; Habel *et al.*, 1998). Finally, the sensor can be pre-packaged in a spot-weldable enclosure for installation on a metallic structure, as shown in Figure 2.11.

When a fibre is surrounded by the host or the buffer material, it is possible that a parasite sensitivity appears in the strain components transverse to the fibre axis. This is particularly true in the case of interferometric sensors (including Bragg-grating sensors), where a transverse pressure will change the fibre's refractive index. This change will be incorrectly interpreted as an axial strain variation. This may be overcome by a proposed method using two superposed Bragg gratings written in a birefringent fibre (Udd *et al.*, 1996) or more simply by shielding the sensor for the transverse strain.

In the case of deformation sensors with a measurement base far longer than the fibre diameter, two installation approaches are possible: full length and local coupling (Inaudi *et al.*, 1994). In the first case, the fibre is in mechanical contact with the host structure along its whole measurement length. In the second case, the fibre is attached to the structure only at the ends of the measurement region and prestressed in-between. The SOFO sensor presented in Figure 2.12 is an example of such an approach (Inaudi *et al.*, 1997; Inaudi, 1997).

The ingress and egress points of the fibres in the structure are other critical details to be considered. These points often represent an important failure source, especially in the case of

Figure 2.11 Packaged sensor for spot-welding installation (courtesy of Roctest).

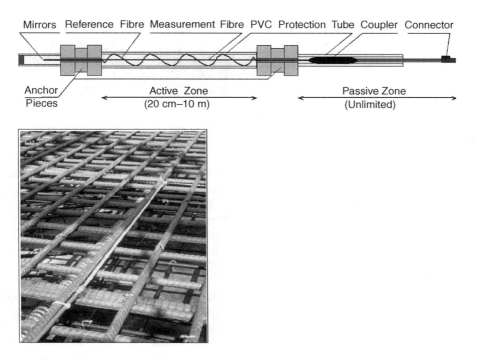

Figure 2.12 Setup and application example of a standard SOFO sensor (courtesy of SMARTEC).

host materials that require external finishing (e.g. removal of the casting forms from a concrete structure or cutting of a composite panel). In the case of concrete structures, these problems can usually be solved by installing appropriate reservation boxes containing and protecting the fibre connectors or splices.

In many cases, the installation of the sensors in the structure constitutes a serious challenge that should not be underestimated. Many trials (and errors!) are often necessary before a reliable and efficient procedure can be established. It is desirable that the installation technique is included from the beginning in the design process, and early trials should be carried out in real conditions, even before the whole sensing system is operational.

Packaged sensors designed for installation in or on civil structures are now on the market. In particular, it is possible to obtain packages to install FBGs and EFPIs on metallic structures or embed then in concrete. Long-gauge sensors based on low-coherence (SOFO system) can also be easily concreted or attached to a surface.

Although the reading unit and software can often be used for monitoring many different types of structure, sensors must be modified and adapted to the host material. Composite materials are generally manufactured in the form of filaments, tapes or sheets, whereas sensors are to be embedded within the structure, depending on the structural layers that have to be monitored. Improper embedding of the sensors may be a source of delamination that causes a significant decrease of mechanical properties. Sensors can also be installed on the surface of the structure, and in this case the optical fibre has to be protected against environmental influences.

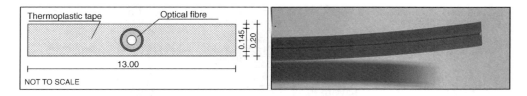

Figure 2.13 Typical cross-section and appearance of a SOFO SMARTape sensor (courtesy of SMARTEC).

On the other hand, if the sensor is designed to monitor strain or deformation, it is necessary to guarantee a good bonding between the optical fibre and the composite. Finally, for an industrial deployment of fibre optic sensors in this domain, it is necessary to package the sensors in a way that makes them as easy to handle as other components used for composite production.

For the monitoring of composites, the solution is found in pre-packaging the measurement optical fibre in a thin composite tape that can then be embedded or surface mounted on the composite structure. The tape gives to the optical fibre the necessary protection against accidental damage during handling and installation. The fibre-reinforced composite tape with integrated optical fibre is called SMARTape, a typical cross-section of which is presented in Figure 2.13. This is a typical example of a distributed coupling strain sensor.

The SMARTape sensing performance was laboratory tested (Glišić and Inaudi, 2003b). In addition, mechanical, microscopic and fatigue tests were performed. The results confirmed the same sensing performance as in case of a standard SOFO sensor and excellent mechanical (robust, resistant, elastic, with no fatigue), thermal (temperature range from −40 to +300 °C) and chemical performance (resistant to aggressive environments).

A SOFO-compatible inclinometer, based on an inverted-pendulum design (Inaudi and Glišić, 2002a), is also available for monitoring structures that are subjected to tilt and is shown in Figure 2.14.

Figure 2.14 SOFO inclinometer (courtesy of SMARTEC).

Figure 2.15   FBG temperature sensor (courtesy of Fibre Sensing) .

For temperature sensing, the sensing element (e.g. an FBG) should be shielded for outside strain and bending, so that it only reacts to temperature. Figure 2.15 shows an FBG temperature sensor. It is also possible to integrate a temperature sensor in the same packaging hosting a strain or deformation sensor. Many different types of packaging for stain sensing with FBGs have been proposed and applied in field-testing and applications (Bugaud *et al.*, 2000).

For the monitoring of relative settlements, an FBG water-levelling system has been proposed. This allows the monitoring of vertical movements with sub-millimetre accuracy (see Figure 2.16).

Fibre-optic accelerometers based on FBGs, shown in Figure 2.17, have also been developed and can be conveniently combined with strain and temperature sensors.

Figure 2.16   FBG water-levelling system (courtesy of SMARTEC).

Figure 2.17   FBG single-axis accelerometer (courtesy of Fibre Sensing).

Figure 2.18   EFPI sensors for strain, pressure and temperature (courtesy of Roctest).

Packaged EFPI sensors and demodulators for civil structural monitoring are commercially available from Roctest and FISO. Figure 2.18 shows examples of EFPI sensors for strain pressure and temperature.

2.4 Distributed Sensing Cables

2.4.1 Introduction

Traditional fibre-optic cable design aims to provide the best possible protection of the fibre itself from any external influence. In particular, it is necessary to shield the optical fibre from external humidity, side pressures, crushing and longitudinal strain applied to the cable. These designs have proven very effective in guaranteeing the longevity of optical fibres used for communications and can be used as sensing elements for monitoring temperatures in the -20 to $+60\,^{\circ}$C range, in conjunction with Brillouin or Raman monitoring systems.

Sensing distributed temperature below $-20\,^{\circ}$C or above $60\,^{\circ}$C requires a specific cable design, especially for Brillouin scattering systems, where it is important to guarantee that the optical fibre does not experience any strain that could be misinterpreted as a temperature change due to the cross-sensitivity between strain and temperature.

On the other hand, the strain sensitivity of Brillouin scattering encourages the use of such systems for distributed strain sensing, in particular to monitor local deformations of large structures such as pipelines, landslides or dams. In these cases, the cable must faithfully transfer the structural strain to the optical fibre, a goal contradicting all experience from telecommunications cable design, where the exact opposite is required.

Finally, when sensing distributed strain it is necessary to measure temperature simultaneously in order to separate the two components. This is usually obtained by installing strain- and temperature-sensing cables in parallel. Therefore, it would be desirable to combine the two functions into a single package.

These very practical requirements have led to the development of cables specifically designed for sensing applications.

All cable designs for distributed sensing share common goals independent of the sensing technique used (Brillouin or Raman scattering) and application domain:

- Optical fibres must be compatible with the sensing system selected: single-mode fibres for Brillouin scattering and (usually) multimode fibres for Raman scattering systems.
- The fibres must be protected from external mechanical actions during installation and while in use. In particular, the cable design must allow easy manipulation without the risk of fibre damage.
- The cable design must allow sufficient shielding of the optical fibres from chemical aggression by water and other harmful substances.
- All optical losses must be kept as low as possible in order not to introduce degradations to the native instrument's distance range.
- Installation of connectors and repair of damaged sensors should be compatible with field operations.

The next section will describe cable designs for high and low temperatures, distributed strain and combined strain and temperature sensing.

2.4.2 Temperature-Sensing Cable

The extreme temperature-sensing cables shown in Figure 2.19 are designed for distributed temperature monitoring over long distances. They consist of up to four single-mode or multimode optical fibres contained in a stainless steel loose tube, protected with stainless steel armouring wires and an optional polymer sheath. These components can be differently combined in order to adapt the cable to the required performance and application. The use of appropriate optical-fibre coating (polyimide or carbon–polyimide) allows the operation over large temperature ranges; the stainless steel protection provides high mechanical and additional chemical resistance, and the polymer sheath guarantees corrosion protection. A carbon coating offers improved resistance to hydrogen darkening. The over-length of the optical fibres is selected in such a way that the fibre is never pulled or compressed, despite the difference in thermal expansion coefficients between glass and steel. The total cable diameter is only 3.8 mm.

These cables can be used in a wide range of applications that require distributed temperature sensing, such as temperature monitoring of concrete in massive structures, waste disposal sites, onshore, off-shore and downhole sites in gas and oil industry, hot spots, cold spots and leakage

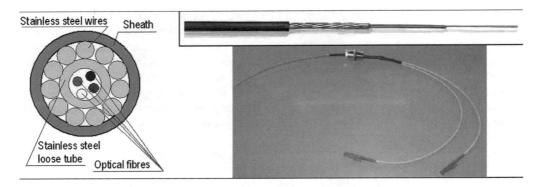

Figure 2.19   Extreme temperature-sensing cable design and termination (courtesy of SMARTEC).

detection of flow lines and reservoirs, fire detection in tunnels and mapping of cryogenic temperatures, just to name a few.

2.4.3 Strain-Sensing Tape: SMARTape

When strain sensing is required, the optical fibre must be bonded to the host material over the whole length. The transfer of strain is to be complete, with no losses due to sliding. Therefore, an excellent bond between the strain-sensing optical fibre and the host structure is to be guaranteed. To allow such a good bond it has been recommended to integrate the optical fibre within a tape in a manner similar to the way reinforcing fibres are integrated in composite materials (Glišić and Inaudi, 2003b). To produce such a tape, a glass-fibre-reinforced thermoplastic with a poly(phenylene sulfide) matrix was selected. This material has excellent mechanical and chemical resistance properties. Since its production involves heating to high temperatures (in order to melt the matrix of the composite material), it is necessary for the fibre to withstand this temperature without damage. In addition, the bonding between the optical fibre coating and the matrix has to be guaranteed. Polyimide-coated optical fibres fit these requirements and, therefore, were selected for this design.

The typical cross-section width of the thermoplastic composite tape used for manufacturing composite structures is in the range of 10–20 mm; therefore, it is not critical for optical-fibre

Figure 2.20   Cross-section picture and picture of the sensing tape: DiTeSt SMARTape (courtesy of SMARTEC).

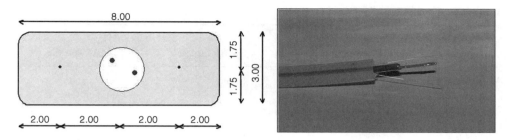

Figure 2.21 SMARTprofile cross-section and sample; the tube contains the free fibres (courtesy of SMARTEC).

integration. However, the thickness of the tape can be as low as 0.2 mm. This dimension is more critical, since the external diameter of a polyimide-coated optical fibre is approximately 0.145 mm. Hence, less than 0.03 mm of tape material remains on the top or bottom of the optical fibre, with the risk that the optical fibre will emerge from the tape. A schematic of the sensing tape cross-section and its appearance is presented in Figure 2.20; the sensing tape's typical dimensions are similar to as given in Figure 2.13.

The use of such sensing tape (called DiTeSt SMARTape) is twofold: it can be used externally, attached to the structure, or embedded between the composite laminates, having also a structural role.

2.4.4 Combined Strain- and Temperature-Sensing: SMARTprofile

The SMARTprofile sensor design combines strain and temperature sensors in a single package. This sensor consists of two bonded and two free single-mode optical fibres embedded in a polyethylene thermoplastic profile. The bonded fibres are used for strain monitoring, while the free fibres are used for temperature measurements and to compensate temperature effects on the bonded fibres. For redundancy, two fibres are included for both strain and temperature monitoring. The profile itself provides good mechanical, chemical and temperature resistance. The size of the profile makes the sensor easy to transport and install by fusing, gluing or clamping. The SMARTprofile (see Figure 2.21) sensor is designed for use in environments often found in civil geotechnical and oil and gas applications (Inaudi and Glišić, 2005). However, this sensor cannot be used in extreme temperature environments or in environments with high chemical pollution. It is not recommended for installation under permanent UV radiation (e.g. sunshine).

Table 2.2 compares the performances of the cable designs presented and specifies their application domains and limitations.

2.5 Software and System Integration

The installation of a permanent monitoring system in a structure brings with it the necessity of managing a large volume of data. Even in a structure with only a few tens of sensors that are measured every hour, the number of data points that are generated can become so large that a manual analysis becomes tedious and very time consuming.

Table 2.2   Comparison of sensor performances and characteristics

	Extreme temperature sensor	SMARTape	SMARTprofile
Measurement parameters	Temperature	Strain	Strain and temperature
Number of fibres	1–4	1	2 for strain and 2 for temperature
Typical loss at 1550 nm (dB km^{-1})	0.3	5	2.0 (for strain) 0.3(for temperature)
Temperature range (°C)	−180 to +300	−180 to +250	−40 to +60
Chemical aggression resistance	Good to excellent (with polymer sheath)	Excellent	Good
Application examples	Fire detection, cryogenic tanks, high-temperature pipeline monitoring, remote heating system monitoring, steam generator monitoring	Pipeline strain, composite pipes monitoring, surface installation on concrete, steel and timber, embedding in composites	Strain, temperature, leakage and third-party intrusion detection for pipelines

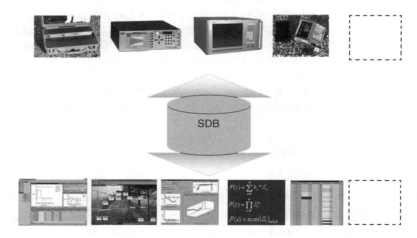

Figure 2.22   Use of a relational database system as an interface layer between DAQ systems and DRA software packages (courtesy of SMARTEC).

Furthermore, the DAQ phase is only the first step in a chain of operations that ultimately leads to high-level information useful for the structure's owner. These analysis steps are increasingly based on the use of software tools that require access to the raw data. This leads to the need for a structured data-storage system that can be accessed by the DAQ software and by the data representation and analysis (DRA) packages.

The natural choice for storing data is of course a database (Inaudi *et al.*, 2002). Since many research groups and companies worldwide are currently developing DAQ and DRA applications, a standard core database helps make these tools more interoperable. For this reason, the core structural database (SDB) structure has been proposed as an open standard for storing structural monitoring data.

The database structure can be used as a central node between the DAQ and the DRA software packages. This allows new measurement systems to be immediately recognized by the existing DRA packages, and vice versa, as shown in Figure 2.22.

2.6 Conclusions and Summary

Fibre-optic structural monitoring is not an isolated and self-sufficient research and industrial field, but rather a way to combine and enhance existing technologies to achieve innovative results.

In the first decade of fibre-optic sensor technology, most efforts were concentrated on the different subsystems. The reading unit and the multiplexing subsystems have seen important developments, and many technologies are today mature for field and industrial applications. Some techniques have emerged, like FBG sensors, low-coherence sensors and EFPIs; others are living a second youth, like intensity-based sensors. New technologies, like Brillouin scattering, have created entirely new sensing possibilities unheard of in the traditional sensing arena. Portable reading units are getting smaller each year and have been successfully operated in demanding environments, like those found in marine and civil engineering applications.

In these last few years, the maturity of the reading unit subsystems has been driven toward the development of reliable sensors and installation techniques. Fibre-optic sensors have been embedded successfully in a number of materials and structures, including composites, concrete, timber and metals. Some of these efforts have led to industrial products that now allow the instrumentation of structures with an increasing number of sensors at reasonable prices. This progress is helped by the continuous development of fibre-optic components like fibres, cables, connectors, couplers and optical switches driven by the much larger telecommunications market.

With structures equipped with hundreds or even thousands of sensors, measuring different parameters each second, the need for automatic data analysis tools is becoming increasingly urgent. Efforts have already been made in this direction. Unfortunately, each type of structure and sensor needs specific processing algorithms. Vibration and modal analysis have attracted much research effort, and geometrical analysis, like curvature measurements, can be easily applied to many types of structure, like tunnels or spatial structures. Many other concepts, like neural networks, fuzzy logic, artificial intelligence, genetic algorithms and data-mining tools, will certainly find an increasing interest for smart processing applications.

The ubiquity of digital networks and cellular communications tools increases the flexibility of the interface subsystems and makes remote sensing not only possible but also economically attractive. Of course, every remote-sensing system has to be based on reliable components, since the need for manual intervention obviously reduces the interest of such systems.

There is more than the developments in each component, however; it is the successful integration of different technologies that is leading to increasingly useful applications. This integration is possible only in highly multidisciplinary teams, including structural, material and sensor engineers.

The final judge of all structural monitoring systems, however, will be the market. Even well-designed and perfectly functioning systems will have to prove their economic interest in order to succeed. Unfortunately, the evaluation of the benefits of a monitoring system is often difficult and the initial additional investments are paid back only in the long term. Furthermore, it is not easy to quantify the benefits of the increased security of one structure or of a better knowledge of its aging characteristics. In many fields, including civil engineering and aeronautics, we are, however, witnessing an investment shift from the construction of new structures to the maintenance and the life-span extension of existing ones. In these domains, fibre-optic sensing certainly has an important role to play.

3

Fibre-Optic Deformation Sensors: Applicability and Interpretation of Measurements

3.1 Strain Components and Strain Time Evolution

3.1.1 Basic Notions

Besides stress, strain is the most important mechanical parameter that reflects structural condition. Taking into account that no real stress-monitoring systems exist for simple and economical on-site applications, strain is the most frequently monitored parameter and strain monitoring is the central topic of this book. Hence, it is important to recall the basic notions relating to the strain as defined in *Strength of Materials* (Brčić, 1989).

Let $A(x_A, y_A, z_A)$ and $B(x_B, y_B, z_B)$ be two points close to each other and belonging to a structural element, where x, y and z denote the Cartesian coordinates of the points in a given local coordinate system. Let L_{A-B} be the length of the segment AB and n is the line defined by the points A and B. If the structural element is subject to a deformation, then points A and B will move to A' and B' respectively, and the length between them will change (elongate or shorten) by ΔL_{A-B}, as shown in Figure 3.1.

The value

$$\varepsilon_n^{\mathrm{av}}(A) = \varepsilon_n^{\mathrm{av}}(x_A, y_A, z_A) = \frac{\Delta L_{A-B}}{L_{A-B}} \tag{3.1}$$

is called the average strain in the proximity of point $A(x_A, y_A, z_A)$ with respect to axis n.

Let $B' \to A'$ along the line n', then the value

$$\varepsilon_n(A) = \varepsilon_n(x_A, y_A, z_A) = \lim_{\substack{B' \to A' \\ n'}} \frac{\Delta L_{A-B}}{L_{A-B}} \tag{3.2}$$

is called the strain at point $A(x_A, y_A, z_A)$ with respect to the axis n. Conventionally, the strain is positive if it indicates elongation (expansion, $\Delta L_{A-B} > 0$) and negative if it indicates shortening

Fibre Optic Methods for Structural Health Monitoring   B. Glišić and D. Inaudi
© 2007 John Wiley & Sons, Ltd

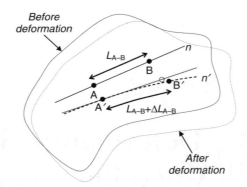

Figure 3.1   Notion of average strain in deformed body.

(contraction, $\Delta L_{A-B} < 0$). The strain is a nondimensional value (real unit: metre/metre can be written, but it is usually omitted) and is often expressed in microstrain ($\mu\varepsilon$) or as a percentage (%), where $1\ \mu\varepsilon = 1\ \mu\text{m m}^{-1} = 10^{-6}\ \text{m m}^{-1} = 0.0001\ \%$.

Let m and n be two coplanar perpendicular axes belonging to a structural element, let A(x_A, y_A, z_A) be their interception point and let M(x_M, y_M, z_M) and N(x_N, y_N, z_N) be points in the proximity of A belonging to the lines m and n respectively. If the structural element is subject to a deformation, then the points A, M and N will move to A′, M′ and N′ respectively, and the angle $\angle$MAN ($= \pi/2$) will change to the angle $\angle$M′A′N′, as presented in Figure 3.2.

The value

$$\gamma_{mn}^{av}(A) = \gamma_{mn}^{av}(x_A, y_A, z_A) = \angle\text{MAN} - \angle\text{M}'\text{A}'\text{N}' = \frac{\pi}{2} - \angle\text{M}'\text{A}'\text{N}' \tag{3.3}$$

is then called the average shear strain in the proximity of point A(x_A, y_A, z_A) with respect to axes m and n.

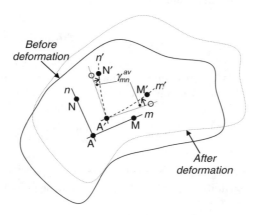

Figure 3.2   Notion of average shear strain in deformed body.

Let $M' \to A'$ along line m' and let $N' \to A'$ along line n', then the value

$$\gamma_{mn}(A) = \gamma_{mn}(x_A, y_A, z_A) = \lim_{\substack{M' \to A', N' \to A' \\ m' \qquad n'}} \left(\frac{\pi}{2} - \angle M'A'N' \right) \tag{3.4}$$

is called the shear strain at point $A(x_A, y_A, z_A)$ with respect to axes m and n. Conventionally, the shear strain is positive if it indicates a decrease of the angle $\angle MAN$ ($\angle MAN = \pi/2 > \angle M'A'N'$) and negative if it indicates its augmentation.

The shear strain is a nondimensional value (real unit: radian can be written, but it is usually omitted) and is often expressed in microradians (μrad) or micrometres per metre (μm m^{-1}), where $1\ \mu$rad $= 10^{-6}$ rad $= 1\ \mu$m m^{-1}.

Since there is an infinity of directions crossing point A, there is also an infinity of strain and shear strain values at point A, depending on the observed directions. However, these values are not mutually independent, but linearly dependent on three independent strain values, called principal (main) strains, defined with respect to three non-coplanar mutually perpendicular axes with the same interception point, called principal (main) axes. More exhaustive considerations on the strain transformations exceed the topic of this book and can be found in handbooks such as the *Strength of Materials* (Brčić, 1989).

The sources of strain in construction materials are manifold (Muravljov, 1989). Strain is most frequently generated by loads (structural strain) and thermal and rheologic effects; but some other sources of strain can also be present, such as residual strain in different materials, endogenous, carbonation and plastic strain in young concrete, alkali-reaction swelling in matured concrete, and so on.

The structural strain is generated by loads and can be elastic (reversible) and plastic (irreversible). The magnitude and the type of the structural strain depends on the load level and on the mechanical properties of the construction materials, such as Young's modulus E, the Poisson coefficient ν and the yield strain ε_y.

Thermal strain is generated by temperature variations in the construction materials and depends on the thermal expansion coefficient α_T. Depending on the boundary conditions, thermal strain can generate stresses (typically in hyperstatic structures), thus it can be coupled with structural strain.

Rheologic strain represents dimensional changes in construction material that occur in time even if no change in load is present (Brčić, 1989). The most common rheologic strains are creep (and relaxation) and shrinkage. Creep and shrinkage can cause significant redistribution of loads in hyperstatic structures and, as in the case of thermal strain, can generate stresses.

Structural and thermal strain occurs in all commonly used construction materials, such as concrete, steel, timber and composites. Usually, only elastic strain is allowed in materials under normal working conditions (service). Rheologic effects affect concrete in particular, where creep and shrinkage can have magnitudes higher than the structural strain (Neville, 1975; CEB-FIP, 1990). These effects can practically be neglected in steel and most composites. The most frequent sources of strain and their relation to different construction materials are schematically presented in Table 3.1 (Muravljov, 1989; Keller, 2003).

The strain components generally occur simultaneously, and the total strain ε_{tot} in the material is equal to the sum of these components:

$$\varepsilon_{\text{tot}} = \varepsilon_E + \varepsilon_P + \varepsilon_T + \varepsilon_\varphi + \varepsilon_{\text{sh}} + \varepsilon_{\text{other}} \tag{3.5}$$

Table 3.1 The most frequent sources of strain for different construction materials

Source of strain	Strain component	Concrete	Steel	Composite	Timber
Load (structural strain)	Elastic strain ε_E	Yes	Yes	Yes	Yes
	Plastic strain ε_P	Yes	Yes	No[b]	Yes
Temperature variations	Thermal strain ε_T	Yes	Yes	Yes	Yes
Rheologic effects	Creep ε_φ	Yes	No[a]	No[a]	Yes
	Shrinkage ε_{sh}	Yes	No	No[c]	No[e]
Others	ε_{other}	Yes	No[d]	No[d]	Yes[e]

[a] Only minimal creep; small percentage of elastic strain (if any).
[b] Depends on reinforcing material; commonly used glass- and carbon-reinforced composites have minimal plastic deformation (if any).
[c] Depends on matrix.
[d] Practically negligible.
[e] Swelling and shrinkage are mainly related to humidity changes.

Depending on the construction material, some strain components in Equation 3.5 may not be present (e.g. shrinkage in the case of steel); that is, they are equal to zero. Simplified guidelines for the evaluation of the above-mentioned strain components are briefly presented in the following subsections. A detailed analysis of all strain components exceeds the topic of this book but can be found in the literature related to specific construction materials.

3.1.2 Elastic and Plastic Structural Strain

Structural strain is present in loaded structures and directly reflects the structural condition. Therefore, in engineering practice, it is very important to be able to evaluate the elastic and plastic structural strain components.

Loads deform the structure and generate strains, stresses and internal forces. The relation between the strain and the stress is governed by the constitutional law of the construction materials. The short-term strain–stress relation in commonly used construction materials is considered as elasto-plastic, linear (steel, composite, simplified analysis of concrete and timber) or nonlinear (more sophisticated analysis of concrete and timber in compression). Examples of linear and nonlinear short-term strain–stress relations are presented in Figure 3.3a and b.

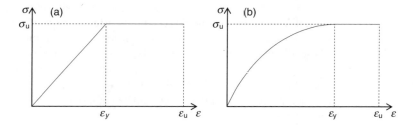

Figure 3.3 Stress–strain relation for elasto-plastic materials in the short term: (a) linear; (b) nonlinear.

Since the construction materials can be considered to be linearly elasto-plastic in the short term, then, except in such cases where a more sophisticated model is necessary, the analysis presented in this book is based on this model. More sophisticated analysis exceeds the topic of this book, but can be carried out following similar principles to those presented here.

A material exhibits linearly elasto-plastic behaviour if the relation between strain and uniaxial stress is given by the following law (Timoshenko and Goodier, 1970):

$$\varepsilon_s < \varepsilon_y \Rightarrow \sigma = \varepsilon_s E; \quad \varepsilon_u \geq \varepsilon_s \geq \varepsilon_y \Rightarrow \sigma = \varepsilon_y E = \sigma_u = \text{const} \tag{3.6}$$

where σ is the stress, ε_s is the structural strain, E is Young modulus, ε_y is the yield strain; σ_u is the ultimate stress and ε_u is the ultimate strain. Equation 3.6 is graphically presented in Figure 3.3a.

Elastic strain (linear or nonlinear) is reversible; that is, the elastic strain is fully recovered if the load is removed. Plastic strain is an irreversible strain; that is, the deformation is permanent if the load is removed. The yield strain represents the limit between elastic and plastic strain in construction materials. In other words, if the strain in a material is lower than the yield strain, then the material is only elastically strained, but if the strain exceeds the yield strain, then the material is both elastically and plastically strained. This is captured in the following:

$$\varepsilon_s < \varepsilon_y \Rightarrow \varepsilon_E = \varepsilon_s; \quad \varepsilon_s \geq \varepsilon_y \Rightarrow \varepsilon_E = \varepsilon_y \wedge \varepsilon_P = \varepsilon_s - \varepsilon_y \tag{3.7}$$

Equations 3.6 and 3.7 are valid for positive stresses and strains (traction). Similar equations can be carried out for negative stress and strain (compression). While in normal service conditions, most structures are operated in the elastic domain, whereas plastic strain is rarely allowed and is usually an indication of damage or insufficient capacity.

Besides the elastic strain generated in the same direction of uniaxial stress, additional elastic strain components are generated in the perpendicular direction and can be expressed as follows for linearly elastic materials (Timoshenko and Goodier, 1970):

$$\varepsilon_{\perp E} = \nu \varepsilon_E = \nu \frac{\sigma}{E} \tag{3.8}$$

where ν is Poisson's coefficient.

The elastic strain distribution in a structural element is not constant and depends on the load intensity, type (forces, distributed forces, moments, etc.) and position, the structural element type (beam, plate, shell, etc.), its geometrical properties (length, cross-section), the material physical properties (homogeneous material or not, isotropic or not), the material mechanical properties (Young's modulus, Poisson's coefficient) and, finally, on the static system of the structure as a whole. Thus, the elastic strain distribution in structural elements can be very complex. A detailed presentation of all the possibilities exceeds the topics of this work and is rather in the domain of separate branches of engineering such as the theory of structures, the theory of plates and shells and the strength of materials. However, the basic notions are summarized below.

The most common structural elements in engineering practice are beams (Đurić and Đurić-Perić, 1990). The most relevant parameter to characterize the structural behaviour of a beam is the elastic strain in the direction perpendicular to its cross-section, namely in directions parallel to the elastic line of the beam (the elastic line connects the cross-sectional centres of gravity). The most frequently used construction materials, such as concrete, steel, timber and

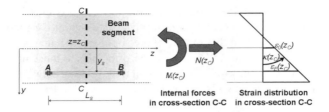

Figure 3.4   Strain distribution in a beam's cross-section (courtesy of SMARTEC).

composite, can be considered as homogeneous at a macro-level, and since the elastic strain we analyse is unidirectional (parallel to elastic line) it is not important whether the material is isotropic or not.

For linear structural elements with constant cross-sections (beams) the strain distribution in the element is given by the following formula (Brčić, 1989):

$$\varepsilon_{\mathrm{E}}(y, z) = \frac{N(z)}{EA} + \frac{M(z)}{EI}\, y = \varepsilon_{0\mathrm{E}}(z) + \kappa_{\mathrm{E}}(z)y \tag{3.9}$$

where z is the local coordinate of the beam along the elastic line, y is the local coordinate of the beam perpendicular to the elastic line, $\varepsilon_{\mathrm{E}}(y, z)$ is the elastic strain at location (y, z); $N(z)$ is the normal (axial) force at location z, E is Young's modulus, A is the cross-sectional area, $\varepsilon_{0\mathrm{E}}(z) = N(z)/EA$ is the central elastic strain (in centre of gravity) at location z, $M(z)$ is the bending moment at location z, I is the moment of inertia and $\kappa_{\mathrm{E}}(z) = M(z)/EI$ is the elastic curvature at location z.

The elastic strain distribution in an arbitrarily chosen cross-section C–C with local coordinate z_{C} depends on the internal forces in this cross-section, the normal (axial) force $N(z_{\mathrm{C}})$ and the bending moment $M(z_{\mathrm{C}})$, and is graphically presented in Figure 3.4.

The strain distribution along a sensor with gauge length L_{s}, installed parallel to the elastic line at distance y_{sen}, as shown in Figure 3.4, depends only on the normal force and the bending moment distributions along the neutral line of the beam and is described by

$$\varepsilon_{\mathrm{E,s}}(z) = \frac{N(z)}{EA} + \frac{M(z)}{EI}\, y_{\mathrm{s}} = \varepsilon_{0\mathrm{E}}(z) + \kappa_{\mathrm{E}}(z)y_{\mathrm{s}} \tag{3.10}$$

The most common loads can be represented in the form of concentrated forces and moments and uniformly distributed forces. In these cases the normal force distribution along beams is constant or linear, whereas the bending moment distribution is constant, linear or parabolic, as presented schematically in the example shown in Figure 3.5. Consequently, the distributions of central strain and curvature along the elastic line are similar to those of the corresponding internal forces, as also presented schematically in Figure 3.5.

It is important to highlight that, in the case of reinforced concrete, the Young's modulus and cross-sectional properties (area and moments of inertia) are not constant in time: Young's modulus evolves in time, whereas the cross-sectional properties are changed by the onset of structural cracks.

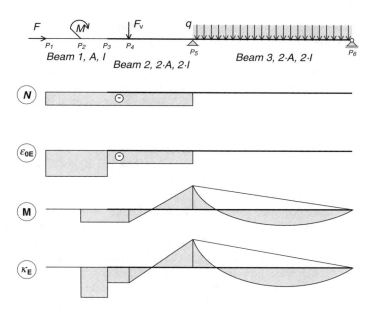

Figure 3.5   Schematic representation of most common loads, corresponding normal force and bending moment distributions, and resulting central strain and curvature distributions.

In anisotropic materials, such as composite or timber, Young's modulus and Poisson's coefficient can be different for different directions.

3.1.3 Thermal Strain

Temperature variations in a structure's environment generate temperature variations in the structure itself. The consequence of these temperature variations are dimensional changes of the construction material in all the three directions, strains and deformations. In common construction materials the following relation between the strain and temperature change is valid (Brčić, 1989):

$$\varepsilon_{n,\mathrm{T}} = \alpha_{\mathrm{T},n} \Delta T \tag{3.11}$$

where ΔT is the temperature change in the construction material, $\alpha_{\mathrm{T},n}$ is the thermal expansion coefficient of the construction material with respect to the axis n and $\varepsilon_{n,\mathrm{T}}$ is the thermal strain with respect to the axis n generated by the temperature change.

In anisotropic materials the thermal expansion coefficient is different for different directions, whereas in isotropic materials the thermal expansion coefficient is identical for all directions. The thermal strain distribution along a sensor depends only on the temperature distributions along the sensor's gauge length L_{s}, and is described by the following equation:

$$\varepsilon_{\mathrm{s},\mathrm{T}}(z) = \alpha_{\mathrm{T},z} \Delta T(z) \tag{3.12}$$

where z is the local coordinate along the sensor's gauge length, ΔT is the temperature change in the construction material along the axis z and $\alpha_{T,z}$ is the thermal expansion coefficient of the construction material with respect to the axis z.

Temperature changes in the structure's environment are not immediately and fully transferred to the whole volume of the structure's construction material. This transfer can be faster or slower depending on the thermal properties of the construction material (thermal capacity and conductivity). That is why thermal gradients are created within the material; consequently, structural strains can also be generated.

For example, lowering the environmental temperature will cause the cooling of the 'skin' of the structure while the 'core' of the material will still have a higher temperature. The skin contraction will be constrained by the core and, thus, a traction strain will be generated in the skin and a contraction in the core.

Once the temperature becomes uniform in the material, the structural strain will disappear in isostatic structures, since they have exactly zero degrees of freedom. However, the structural strain generated by thermal strain will still be present in hyperstatic structures.

In order to determine the thermal strain and to be able to uncouple it from structural strain it is necessary to monitor temperature variations. A lack of data concerning temperature changes can cause significant errors in the data analysis. Typical values of thermal expansion coefficients of steel and concrete are as follows (Muravljov, 1989):

$$\alpha_{T,steel} = 10\text{--}12\,\mu\varepsilon\,{}^{\circ}C^{-1}$$

$$\alpha_{T,concrete} = 8\text{--}12\,\mu\varepsilon\,{}^{\circ}C^{-1}$$

Taking into account the typical values of the Young's moduli for steel and concrete ($E_{steel} = 210$ GPa, $E_{concrete} = 28$ GPa), a temperature variation of $\Delta T = 10\,{}^{\circ}C$ creates strain changes equivalent to significant stress changes, as shown in Equations 3.13 and 3.14:

$$\text{Steel}: \quad \Delta T = 10^{\circ}C \Rightarrow \varepsilon_{T,steel} = 100\text{--}120\,\mu\varepsilon \sim 21\text{--}25\,\text{MPa} \tag{3.13}$$

$$\text{Concrete}: \quad \Delta T = 10\,{}^{\circ}C \Rightarrow \varepsilon_{T,steel} = 80\text{--}120\,\mu\varepsilon \sim 2\text{--}3\,\text{MPa} \tag{3.14}$$

The stress values presented in the above equations actually represent the error in the estimation of the stresses if the temperature changes (and consequently the thermal strain) were not measured or neglected in the analysis. This error is several times higher in geographic zones where seasonal temperature variations are bigger; for example, typical maximal temperature variations in middle Europe range between -20 and $+40\,{}^{\circ}C$ ($\Delta T_{max} = 60\,{}^{\circ}C$).

In the structural analysis and in the analysis of monitoring data, the thermal expansion coefficient is commonly assumed to be constant. However, the thermal expansion coefficient can depend on temperature magnitude and the humidity of material (Christensen, 1991) (concrete, timber, composites), which may induce errors in the data analysis.

3.1.4 Creep

Creep is a time-dependent dimensional change of structurally strained (loaded) materials. Creep develops slowly, and its final value is usually achieved only after several years. The

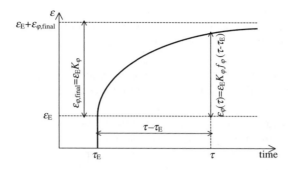

Figure 3.6   Graphical representation of creep evolution.

elasto-plastic materials that feature creep are called viscoelasto-plastic materials. A typical example of a viscoelasto-plastic material is concrete.

Assuming that a constant elastic strain is generated in a construction material, a simplified equation for creep evolution is given by (CEB-FIP, 1990)

$$\varepsilon_\varphi(\tau) = \varepsilon_E K_\varphi f_\varphi(\tau - \tau_E) \tag{3.15}$$

where τ is time ($\tau > \tau_E$), τ_E is the time of loading (structural straining), $\varepsilon_\varphi(\tau)$ is the creep at time τ; ε_E is the elastic strain, K_φ is the creep coefficient and $f_\varphi(\tau - \tau_E)$ is the creep function ($f_\varphi(0) = 0, f_\varphi(\tau \to \infty) \to 1$).

Equation 3.15 is represented graphically in Figure 3.6.

Creep exhibits the following important properties:

1. The final creep value is proportional to the elastic strain
2. The value of the creep at an arbitrary time τ depends on the time of application of elastic strain τ_E
3. If there are several increments of elastic strain generated in different times, then the creep at an arbitrary time τ is equal to the sum of creeps that would be generated by each increment separately
4. Creep is reversible; that is, if the material is unstrained then the creep will eventually return to zero
5. In the case of steel and most composites, the creep coefficient is very small (small percentage of elastic strain) and in most the cases can be neglected
6. In the case of concrete, the creep coefficient depends on the concrete maturity; it ranges between 1 and 3 for mature concrete (28 days) and can be as big as 5.6 for concrete loaded 1 day after the pouring
7. In the case of concrete, creep also depends on relative humidity, structural member size and shape, and exposure to environment.

The consequence of creep in an isostatic structure is an increase of the global deformation of the structure without generating parasitic structural strain and stress. In the case of hyperstatic structures, creep can cause significant strain and stress redistributions.

The creep magnitude can be very high (higher than the magnitude of elastic strain), which is why it is very important to evaluate it during long-term monitoring. Creep cannot be monitored directly; therefore, the following two alternatives are proposed:

1. Evaluate creep on specimens built of the same construction material
2. Use approximate numerical models proposed in the literature.

The first alternative can be expensive (representative number of specimens needed, controlled conditions, long-term monitoring) and time consuming (several years of maintenance and monitoring needed); hence, it is applied only for very particular structures. However, this alternative is more accurate, since it provides real, quantitative data.

The second option is inexpensive and relatively easy to perform, but it is less accurate since a general model is applied that may not be fully representative of the real construction material. Nevertheless, the accuracy obtained is still acceptable and the use of this alternative is justified in most cases.

In both cases it is important to observe and record the load history, if an evaluation of creep is required.

3.1.5 Shrinkage

Shrinkage is a slow, time-dependent dimensional change of nonloaded materials. It occurs in all three dimensions and is an intrinsic property of materials. Shrinkage develops slowly and its final value is usually achieved after several years. A simplified equation for the shrinkage evolution is given by (CEB-FIP, 1990)

$$\varepsilon_{sh}(\tau) = \varepsilon_{sh,\,final}\, f_{sh}(\tau - \tau_{c0}) \tag{3.16}$$

where τ is time ($\tau > \tau_{c0}$), τ_{c0} is the time when the shrinkage is initiated (in case of concrete this is time of pouring), $\varepsilon_{sh}(\tau)$ is the shrinkage at time τ; $\varepsilon_{sh,\,final}$ is the final value of shrinkage and $f_{sh}(\tau - \tau_{c0})$ is the shrinkage function ($f_{sh}(0) = 0$, $f_{sh}(\tau \to \infty) \to 1$).

Equation 3.16 is represented graphically in Figure 3.7.

Shrinkage exhibits the following important properties:

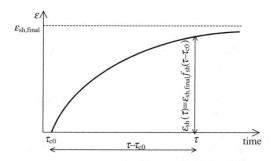

Figure 3.7   Graphical representation of shrinkage evolution.

1. Shrinkage is an intrinsic property of a material and it is independent of load
2. The value of the shrinkage at an arbitrary time τ depends on the time of shrinkage initiation τ_{c0}; in the case of concrete, τ_{c0} corresponds to the time of pouring
3. Shrinkage is, in general, irreversible (can be reversible only under some particular environmental conditions)
4. Steel and most composites (except in the short term) are not subject to shrinkage
5. Concrete is subject to shrinkage, and final values can reach 200–530 $\mu\varepsilon$
6. In the case of concrete, shrinkage also depends on the relative humidity, structural member size and shape, and the exposure to the environment.

The consequence of shrinkage in an isostatic structure is a change of the global deformation of the structure without generating parasitic structural strain and stress. In the case of hyperstatic structures, the shrinkage can cause significant strain and stress redistribution.

The shrinkage magnitude can be very important compared with the magnitude of elastic strain; therefore, it is very important to evaluate it during monitoring. The issues in shrinkage monitoring are similar to those related to creep (see Section 3.1.4) and, therefore, will not be repeated here.

3.1.6 Reference Time and Reference Measurement

In civil structures, all the strain components vary over time. Consequently, different strain components can have different values at different times. Monitoring is a record of the total strain evolution in time. The strain components presented in previous sections in general occur simultaneously; consequently, the total strain ε_{tot} in a material observed at time τ consists of a sum of all the components:

$$\varepsilon_{tot}(\tau) = \varepsilon_E(\tau) + \varepsilon_P(\tau) + \varepsilon_T(\tau) + \varepsilon_\varphi(\tau) + \varepsilon_{sh}(\tau) + \varepsilon_{other}(\tau) \qquad (3.17)$$

In engineering practice, strain evolution is monitored starting at one particular moment in time, called the reference time (e.g. the moment of installation of the monitoring system, pouring of concrete, end of construction work, etc.). The earliest possible reference time is actually the moment of installation of the sensor on the structure being monitored. Let τ_r be the reference time; then:

$$\varepsilon_{Q,\tau-\tau_r} = \varepsilon_Q(\tau) - \varepsilon_Q(\tau_r) \qquad (3.18)$$

where Q is the strain component index ($Q \in \{$'P', 'E', 'φ', 'sh', 'other'$\}$), $\varepsilon_{Q,\tau-\tau_r}$ is the strain component change between times τ_r and τ, $\varepsilon_Q(\tau)$ is the strain component at time $\tau > \tau_r$ and $\varepsilon_Q(\tau_r)$ is the strain component at the reference time.

The sensors cannot distinguish between the different strain components and always measure the total strain. Combining Equations 3.17 and 3.18, the following expression is obtained:

$$\varepsilon_{tot,\tau-\tau_r} = \varepsilon_{tot}(\tau) - \varepsilon_{tot}(\tau_r) \qquad (3.19)$$

The sensor reading performed at the reference time is conventionally called the reference reading, the reference measurement or the zero measurement. The sensor cannot determine the strain developed in the past (i.e. before it was installed on the structure). Consequently, the reading of a sensor is, in general, not equal to the actual strain in the structure. In other words, the reference reading in general does not represent the unstrained state of the structure, except if the sensor was installed at the time of construction (e.g. embedding in the concrete) or is installed on an unstrained structural element. This statement is described by the following expression:

$$\varepsilon_m(\tau_r) \neq \varepsilon_{tot}(\tau_r); \quad \varepsilon_m(\tau) \neq \varepsilon_{tot}(\tau) \tag{3.20}$$

where $\varepsilon_m(\tau)$ is the sensor reading at time τ and $\varepsilon_m(\tau_r)$ is the reference reading (i.e. sensor reading at time τ_r).

The sensor measures strain changes relative to the reference reading. Thus, Equation 3.19 can be expanded to

$$\varepsilon_{m,\tau-\tau_r} = \varepsilon_m(\tau) - \varepsilon_m(\tau_r) = \varepsilon_{tot,\tau-\tau_r} = \varepsilon_{tot}(\tau) - \varepsilon_{tot}(\tau_r) \tag{3.21}$$

where $\varepsilon_{m,\tau-\tau_r}$ is the total strain change between times τ_r and τ measured (registered, read) with the sensor.

Finally, let $t = \tau - \tau_r$ be the time elapsed from the reference time τ_r to an observed moment in time τ. Then, Equation 3.21 transforms to

$$\varepsilon_{m,t} = \varepsilon_m(\tau_r + t) - \varepsilon_m(\tau_r) = \varepsilon_{tot.t} = \varepsilon_{tot}(\tau_r + t) - \varepsilon_{tot}(\tau_r) \tag{3.22}$$

where $\varepsilon_{m,t}$ is the total strain change measured (registered, read) with the sensor, with respect to reference reading and after the time t elapsed from reference time τ_r (i.e. total strain change at time $\tau = \tau_r + t$ with respect to strain at time τ_r).

Equation 3.22 is of particular importance and is the basic expression used in the following sections and chapters. In order to make reading of the text easier, the time t may intentionally be omitted in the further text. However, if not otherwise stated, then it should always be considered as present (i.e. it may be written ε_Q but it should be understood as $\varepsilon_{Q,t}$ if not otherwise stated).

3.2 Sensor Gauge Length and Measurement

3.2.1 Introduction

With reference to their spatial disposition, sensors are classified as discrete or point sensors or as continuous or distributed sensors. A point sensor measures a parameter related to a single position in the structure, whereas distributed sensors measure the parameter at several positions and can practically replace a chain of point sensors.

Point sensors designed to measure relative displacement or average strain between two predefined points of a structure are called deformation sensors. The distance between these two points is called the gauge length of the sensor. With respect to the gauge length, the sensors are conventionally classified in two groups: short-gauge and long-gauge sensors. Traditional sensors, such as strain gauges and vibrating wires, belong to the group of short-gauge sensors. Depending on their type and packaging, optical-fibre sensors can function as short-gauge or as long-gauge sensors.

The availability of long-gauge fibre-optic sensors has opened new and interesting possibilities for structural monitoring. Long-gauge sensors allow the measurement of deformations over measurement bases that can reach tens of metres with resolutions in the micrometre range (see Chapter 2). The long-gauge sensor theory can also be applied to distributed sensing systems (see Section 3.2.7).

Using long-gauge sensors, it is possible to cover the whole volume of a structure with sensors, therefore enabling a global monitoring of it. This constitutes a fundamental departure from standard practice, which is based on the choice of a reduced number of points supposed to be representative of the whole structural behaviour and their instrumentation with short-gauge sensors. This common approach will give interesting information on the local behaviour of the construction materials, but might miss behaviours and degradations that occur at locations that are not instrumented. On the contrary, long-gauge sensors allow the monitoring of a structure as a whole, so that any phenomenon that has an impact on the global structural behaviour is detected and quantified.

Basic notions concerning the gauge length of a deformation sensor and its influence on the measurements obtained are presented below.

3.2.2 Deformation Sensor Measurements

Frequently used construction materials, and notably concrete, can be affected by local defects, such as cracks, air pockets and inclusions. All these defects introduce discontinuities in the mechanical material properties at a meso-level. More indicative for structural behaviour, however, are material properties at the macro-level. For example, reinforced concrete structures are mainly analysed as built of a homogenous material – cracked reinforced concrete. Therefore, for structural monitoring purposes it is necessary to use sensors that are insensitive to material discontinuities at the micro- and meso-levels.

In inhomogeneous materials, the gauge length of a deformation sensor can cross several discontinuities that influence the measurement and its interpretation. A description of the measurements performed by a deformation sensor is presented in Figure 3.8 and Equation 3.23 (Glišić and Inaudi, 2002a).

If A and B are the sensor anchoring points as shown in Figure 3.8, then the measurement of the sensor represents a relative displacement between them. The measurement of the sensor is

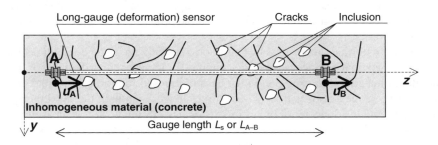

Figure 3.8   Schema of a long-gauge sensor installed on an inhomogeneous material (e.g. concrete) (courtesy of SMARTEC).

then expressed as follows:

$$m_{s,t} = \Delta L_{A-B,t} = u_{B,t} - u_{A,t} = \int_{z_A}^{z_B} \varepsilon_{tot,t}(z)\,dz + \sum_{i=n_{c,A}}^{n_{c,B}} \Delta w_{c,i,t} + \sum_{j=n_{incl,A}}^{n_{incl,B}} \Delta w_{incl,j,t} \quad (3.23)$$

where t is the time elapsed with respect to a reference time τ_r, $m_{s,t}$ is the change of deformation measured by a long-gauge sensor after a time t, $\Delta L_{A-B,t}$ is the change in distance between points A and B (elongation or shortening) after a time t, $u_{A,t}$ and $u_{B,t}$ are the changes in displacements of points A and B respectively in the direction of the sensor after a time t, z_A and z_B are the coordinates of points A and B respectively, $\varepsilon_{tot,t}(z)$ is the change in total strain in the material in the direction of the sensor at location z and at time t, i is the number of cracks crossing the sensor between points A and B, $n_{c,A}$ and $n_{c,B}$ are the numbers of the first crack (closest to point A) and the last crack (closest to point B) included between points A and B, $\Delta w_{c,i,t}$ is the change in size of crack i in the direction of the sensor at time t, j is the number of inclusions crossing the sensor between points A and B, $n_{incl,A}$ and $n_{incl,B}$ are the numbers of the first inclusion (closest to point A) and the last inclusion (closest to point B) included between points A and B and $\Delta w_{incl,j,t}$ is the dimensional change of inclusion j in the direction of the sensor at time t.

Since the deformation sensor measures the relative displacement between two points in a structure, the measurement represents the integral of strain over the sensor's length added to the sum of all discontinuities crossed (cracks and/or inclusions) dimensional changes (see Equation 3.1). Finally, the average strain over the length of the sensor is calculated as the ratio between the measured relative displacement and the gauge length, as presented in Equation 3.24:

$$\varepsilon_{m,t} = \frac{m_{s,t}}{L_s} \quad (3.24)$$

where $L_s = L_{A-B}$ is the gauge length (distance between points A and B) and $\varepsilon_{m,t}$ is the measured change in average strain in the material over the gauge length of the sensor at time t.

A short-gauge deformation sensor has a gauge length shorter than the distance between two discontinuities or comparable to the dimensions of the inclusions in the material monitored. Therefore, the measurement performed with short-gauge sensors is strongly influenced by local defects; it provides information related to local material properties and is not suitable for global structural monitoring.

A long-gauge deformation sensor is by definition a sensor with a gauge-length several times longer than the maximal distance between discontinuities or the maximal diameter of inclusions in a monitored material. For example, in the case of cracked reinforced concrete, the gauge length of a long-gauge sensor is to be several times longer than both the maximum distance between cracks and the diameter of inclusions. The main advantage of this measurement is in its nature: since it is obtained by averaging the strain over long measurement basis, it is not influenced by local material discontinuities and inclusions. Thus, the measurement contains information related to global structural behaviour rather than the local material behaviour. This concept is developed further in Section 3.2.3.

3.2.3 Global Structural Monitoring: Basic Notions

Monitored parameters can be observed at the material or at the structural level. The main difference between these two levels is in the monitoring strategy and monitoring system used: material monitoring provides useful information related to the material behaviour, but reduced information concerning the structural behaviour; structural monitoring provides better information related to the structural behaviour. For inhomogeneous materials, structural monitoring can only be performed using long-gauge sensors, as demonstrated by the following simplified example (Glišić and Inaudi, 2002a).

Let us consider a concrete beam subject to short-term bending moments M at its extremities as shown in Figure 3.9. Consequently, the most relevant parameters to be monitored at the structural level are the curvature κ_E and radius of curvature r_E. The relation between the bending moment M, the curvature κ_E and radius r_E of curvature is given by

$$\kappa_{E,t} = \frac{1}{r_{E,t}} = \frac{M_t}{E_{eq,t} I_{eq,t}} \qquad (3.25)$$

where $\kappa_{E,t}$ is the change in curvature at time t, M_t is the change in bending moment at time t, $r_{E,t}$ is the change in radius of curvature at time t and $E_{eq,t} I_{eq,t}$ is the equivalent cross-sectional stiffness of the cracked concrete element (equivalent Young's modulus of elasticity multiplied by equivalent moment of inertia) at time t.

To monitor curvature, it is necessary to use at least two sensors installed in the same cross-section but at different distances from the neutral axis. By preference, one sensor is installed in the compressed area of concrete (i.e. on the top of the cross-section) and the other sensor is installed in the tensioned (and cracked) area of the concrete (i.e. at the bottom of the cross-section).

Let 1–1 and 2–2 be two arbitrary selected cross-sections instrumented with short-gauge sensors. The notation for these sensors is 'g' with indexes 't' for top, 'b' for bottom, and '1' and '2' for sections 1–1 and 2–2 respectively (see Figure 3.9). Let us consider two parallel long-gauge sensors (notation 's' in Figure 3.9) installed on the top and bottom of the cross-sections as shown in Figure 3.9. Finally, let all the sensors be installed before the element

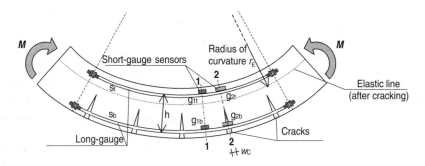

Figure 3.9 Comparison between long-gauge and short-gauge sensor measurements on a bent concrete element (courtesy of SMARTEC).

is loaded (i.e. before the cracks have occurred). In the case of a homogeneous material, the equation to calculate the curvature and the radius of curvature is

$$\kappa_{E,t} = \frac{1}{r_{E,t}} = \frac{\varepsilon_{b,t} - \varepsilon_{t,t}}{h} \tag{3.26}$$

where $\varepsilon_{t,t}$ and $\varepsilon_{b,t}$ are the changes in strains monitored at the top and at the bottom of the cross-section respectively, using the corresponding pair of sensors (on the top and at the bottom); h is the distance between the top and bottom sensors.

After the load is applied to the concrete, the element bends with radius of curvature r_E, the cracks occur with an approximate opening of Δw_c and the elastic line (neutral axes in the cross-sections) moves toward the compressed area of concrete. As an extreme case, let one crack occur exactly on the short-gauge sensor g_{2b} and let the other crack occur in a manner that the short-gauge sensor g_{1b} is exactly in the centre between two cracks (see Figure 3.9).

All the sensors (short- and long-gauge included) installed on the top of the cross-section are in the compressed area of concrete, placed at the same distance from the neutral axis and, consequently, measure the same value of the strain. Thus, the following equation is practically valid:

$$\varepsilon_{st,t} = \varepsilon_{g_{1t},t} = \varepsilon_{g_{2t},t} \tag{3.27}$$

In contrast, the sensors installed at the bottom of the cross-sections measure mutually different values: the short-gauge sensor g_{1b} measures a strain in the tensioned part of the concrete, (i.e. very small value in terms of microstrain) and the short-gauge sensor g_{2b} measures the ratio between the crack opening and gauge length (i.e. extremely big value in terms of microstrain), if it was not broken due to crack opening. The long-gauge sensor will measure an average strain with a value taking into account the variations of strain in the concrete and the crack openings. The following expression is valid:

$$\varepsilon_{g_{2b},t} \approx \frac{\Delta w_{c,t}}{L_{g2}} > \varepsilon_{Sb,t} > \varepsilon_{g_{1b},t} \approx 0 \tag{3.28}$$

The short-gauge sensor g_{1b} provides information on the local material level (strain in tensioned concrete), but along with short-gauge sensor g_{1t} cannot be used to evaluate the radius of curvature (i.e. cannot be used for monitoring at the structural level); Equation 3.26 applied with the short gauge sensors installed in section 1–1 will give underestimated values of the curvature.

The short-gauge sensor g_{2b} also provides information on local material level (related to the crack opening), but also cannot be used along with short-gauge sensor g_{2t} to evaluate the radius of curvature (i.e. cannot be used for monitoring at the structural level); Equation 3.26 applied with the short-gauge sensors installed in section 2–2 will give a greatly overestimated value of the curvature.

However, Equation 3.26 is still valid in the case of long-gauge sensors; that is, the information that they provide is relevant for monitoring at structural level. Thus, the long-gauge sensors follow by their nature the behaviour of concrete, while fully respecting the philosophy of the reinforced concrete as a homogeneous material at the macro-level.

Short-gauge sensors can be used for structural monitoring in homogeneous materials, but the use of the long-gauge sensors is also recommended. Having a longer measurement basis, the long-gauge sensors monitor bigger areas in the structure and, consequently, the probability of detecting a malfunction, damage or critical strain is increased.

The analysis of sensor measurements depending on the gauge length is presented in Sections 3.2.4 and 3.2.5.

3.2.4 Sensor Measurement Dependence on Strain Distribution: Maximal Gauge Length

A long-gauge sensor is insensitive to material discontinuities and measures an average strain, as presented in Equation 3.23. However, the gauge length of the sensor must be carefully selected in order to avoid other possible pitfalls. General guidelines for the determination of the sensor's maximal gauge length depending on the expected strain distribution in loaded structural elements are presented in this section.

Let us now observe a segment of a beam element with a constant cross-section having an arbitrary strain distribution along the sensor, as shown in the left-hand side of Figure 3.10, and equipped with a long-gauge sensor.

The deformation sensor measures the average strain, as presented in Equations 3.23 and 3.24 and in the centre of Figure 3.10. Since the sensors involve a relatively long length (can be several metres long), the average strain value is most commonly attributed to the position at the sensor's centre, as shown in the right-hand side of Figure 3.10 (point C). This obviously involves a certain error in the measurement (see Figure 3.10).

The absolute error in the measurement caused by a long gauge length and attributed to the centre of the sensor is given by

$$\delta\varepsilon_{C,t} = \varepsilon_{m,t} - \varepsilon_{C,t} \qquad (3.29a)$$

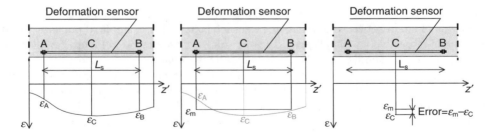

Figure 3.10 Comparison between the real strain distribution and average strain measured with a long-gauge sensor.

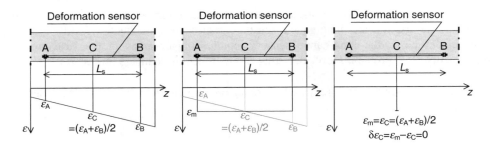

Figure 3.11   In the case of a constant or linear distribution of strain, the average strain measured by a long-gauge sensor is equal to the strain in the middle of the sensor.

and the relative error is calculated using

$$\delta^* \varepsilon_{C,t} = \frac{\delta \varepsilon_{C,t}}{\varepsilon_{C,t}} = \frac{\varepsilon_{m,t}}{\varepsilon_{C,t}} - 1 \qquad (3.29b)$$

The estimation of the error is, in general, difficult, since the real strain distribution along the sensor is not known. However, relatively accurate estimations can be made, based on assumptions taken from the theory of structures.

In the case that neither a horizontal or vertical load (distributed or concentrated) nor a concentrated moment is present between the points A and B, the distribution of strain between them is constant or linear. For example if the sensor is installed within any of the segments P_1-P_2, P_2-P_3, P_3-P_4, or P_4-P_5 presented in Figure 3.5, the strain distribution between the points A *and* B of the sensor will be constant or linear.

For a constant or linear distribution of strain along the sensor's gauge length, the measured average strain value is equal to the real strain value at the centre of the sensor. This fact is illustrated in Figure 3.11 and can easily be proved using Equations 3.23, 3.24 and 3.10. Consequently, the absolute error in the measurement presented in Equations 3.29a and b is zero and the length of the sensor does not influence the measurement results.

If a concentrated force is the only vertical load present between points A and B, then the distribution of strain between them has a broken-line pattern. For example, if points A and B are installed respectively within segments P_3-P_4 and P_4-P_5 presented in Figure 3.5, then the strain will have a broken-line distribution between points A and B of the sensor.

For a 'broken-line' distribution (see Figure 3.12) of strain along the sensor's gauge length, the measured average strain is different from the value of the strain in the centre of the sensor. The worst case occurs when the centre of the sensor coincides with the 'breaking' point. The estimation of error for the worst case is presented graphically in Figure 3.12 and can be proved using Equations 3.23, 3.24 and 3.10.

The absolute and relative errors in measurement determined in Figure 3.12 are expressed by the following equations:

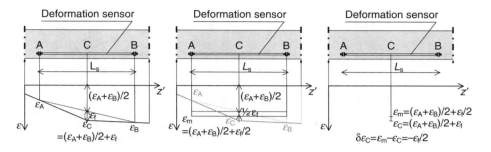

Figure 3.12 In the case of a 'broken-line' distribution of strain, the average strain measured by a long-gauge sensor is lower in absolute value than the strain in the middle of the sensor.

$$\delta\varepsilon_{C,t} = \varepsilon_{m,t} - \varepsilon_{C,t} = -\frac{1}{2}\,\varepsilon_{f,t} \tag{3.30a}$$

$$\delta^{*}\varepsilon_{C,t} = -\frac{1}{2}\frac{\varepsilon_{f,t}}{\varepsilon_{C,t}} \tag{3.30b}$$

where

$$\varepsilon_{f,t} = \varepsilon_{C,t} - \frac{1}{2}(\varepsilon_{A,t} + \varepsilon_{B,t}) \tag{3.31}$$

The errors presented in Equations 3.30a and 3.30b a depend on the sensor's gauge length, as shown in Equation 3.32 (see Figure 3.12 and Equation 3.31):

$$L_{s} \to 0 \Rightarrow \varepsilon_{A,t} \to \varepsilon_{B,t} \to \varepsilon_{C,t} \Rightarrow \varepsilon_{f,t} \to 0 \Rightarrow \delta\varepsilon_{C,t} \to 0 \Rightarrow \delta^{*}\varepsilon_{C,t} \to 0 \tag{3.32}$$

Let L_{Beam} be the length of the beam with a broken-line distribution of strain along the lines parallel to the elastic line (e.g. part of beam 2 in Figure 3.5) and let L_s be the gauge length of the sensor installed parallel to the elastic line. Then, the errors in measurement can be calculated using

$$\delta\varepsilon_{C,t} = -\frac{1}{2}\,\varepsilon_{f,L_{Beam},t}\,\frac{L_{s}}{L_{Beam}} \tag{3.33a}$$

$$\delta^{*}\varepsilon_{C,t} = -\frac{1}{2}\frac{\varepsilon_{f,L_{Beam},t}}{\varepsilon_{C,t}}\frac{L_{s}}{L_{Beam}} \tag{3.33b}$$

where $\varepsilon_{f,L_{Beam},t}$ is the value determined using Equation 3.31 for points A_{Beam} and B_{Beam} set at extremities and 'breaking' point C_{Beam} set in the middle of the beam.

Equations 3.33a and 3.33b demonstrate that that the measurement errors decrease proportionally to the gauge length (shorter sensor, less error). Equations 3.35 and 3.36 can be used to determine the sensor's gauge length to provide measurements within acceptable error limits.

If the desired accuracy is not achievable using a single sensor, then it is advisable to consider the use of two sensors installed left and right with respect to the breaking point. The breaking

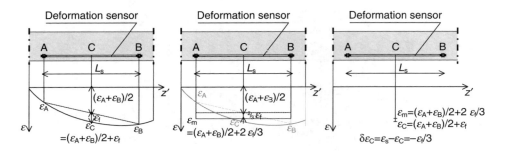

Figure 3.13 In the case of a parabolic strain distribution, the average strain measured by a long-gauge sensor is lower in absolute value than the strain in the middle of the sensor.

point is usually easy to identify since it coincides with the position where a concentrated force is applied to the beam (external load, abutment, support, etc.).

In the case where a uniformly distributed force is the only vertical load present between points A and B, the distribution of strain between them is parabolic. For example, if the sensor is installed within the segment P_5–P_6 presented in Figure 3.5, then the strain will have parabolic distribution between points A and B of the sensor.

For a parabolic distribution of strain along the sensor's gauge length, the measured average strain is different from the value of the strain at the centre of the sensor. The estimation of the error is presented graphically in Figure 3.13 and can be demonstrated using Equations 3.23, 3.24 and 3.10.

The errors in the measurements presented in Figure 3.13 are expressed by the following equations:

$$\delta \varepsilon_{C,t} = \varepsilon_{m,t} - \varepsilon_{C,t} = -\frac{1}{3}\varepsilon_{f,t} \tag{3.34a}$$

$$\delta^* \varepsilon_{C,t} = -\frac{1}{3}\frac{\varepsilon_{f,t}}{\varepsilon_{C,t}} \tag{3.34b}$$

where

$$\varepsilon_{f,t} = \varepsilon_{C,t} - \frac{1}{2}(\varepsilon_{A,t} + \varepsilon_{B,t}) \tag{3.35}$$

as in the case of the broken-line distribution (see Equation 3.31).

The errors presented in Equations 3.34a and 3.34b depend on the sensor's gauge length, as shown for the broken-line case in Equation 3.32 (see Figure 3.13 and Equation 3.35).

Let L_{Beam} be the length of the beam with a parabolic distribution of strain along the lines parallel to the elastic line (e.g. beam 3 in Figure 3.5) and let L_s be the gauge length of the sensor installed parallel to the elastic line. Then the errors in the measurement can be calculated using the following formulas:

$$\delta \varepsilon_{C,t} = -\frac{1}{3}\varepsilon_{f,L_{Beam},t}\frac{L_s^2}{L_{Beam}^2} \tag{3.36a}$$

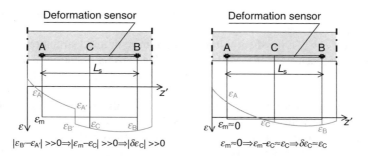

Figure 3.14 Analysis of measurement error in the case of a strain distribution discontinuity (left-hand side) and the existence of inflection (right-hand side) within the sensor's gauge length.

$$\delta^* \varepsilon_{C,t} = -\frac{1}{3} \frac{\varepsilon_{f,L_{Beam},t}}{\varepsilon_{C,t}} \frac{L_s^2}{L_{Beam}^2} \qquad (3.36b)$$

where $\varepsilon_{f,L_{Beam},t}$ is the value determined using Equation 3.35 for points A_{Beam} and B_{Beam} set at extremities and point C_{Beam} set in the middle of the beam (similar to a broken-line distribution).

Equations 3.36a and 3.36b demonstrate that the error of the measurement decreases with the square of the gauge length (shorter sensor, less error). Equations 3.36a and 3.36b can be used to determine a sensor's gauge length to provide measurement within acceptable error limits.

The strain distribution function may often have some exceptional points, such as a discontinuity point ('jump' in the diagram due to a concentrated external moment load or change in the cross-section stiffness, see points P_2 and P_3 in Figure 3.5) or zero point (e.g. inflection point), as presented in Figure 3.14.

These cases are to be studied carefully with respect to the monitoring aims before the final decision concerning the positions of sensors is taken. In the case of a strain distribution discontinuity, the absolute error might be very high, depending on the magnitude of the discontinuity ('jump' in diagram) as presented in the left-hand side of Figure 3.14 and in the following expression:

$$|\varepsilon_{B'} - \varepsilon_{A'}| \gg 0 \Rightarrow |\varepsilon_m - \varepsilon_C| \gg 0 \Rightarrow |\delta \varepsilon_C| \gg 0 \qquad (3.37)$$

where $|\varepsilon_{B'} - \varepsilon_{A'}|$ represents the magnitude of strain discontinuity.

The inflection point influence is presented in the right-hand side diagram of Figure 3.14 and Equation 3.38. The absolute error may have a magnitude close to the measured value (i.e. the relative error may be close to 1.00, which is obviously not acceptable).

$$\varepsilon_m \approx 0 \Rightarrow \varepsilon_m - \varepsilon_C \approx \varepsilon_C \Rightarrow \delta \varepsilon_C \approx \varepsilon_C \Rightarrow \delta^* \varepsilon_C \approx 1.00 \qquad (3.38)$$

For the previously presented cases, if the desired accuracy is not achievable using a single sensor, then it is advisable to consider the use of two sensors installed on each side of an exceptional point, having provided that the identification of such a point location can be accurately determined. An additional sensor can be used to quantify the influence of the discontinuity itself. The decision about what strategy is to be used (single sensor or two sensors) must take into account the monitoring aims.

In a general case, when the distribution of strain in monitored structure follows a known law, the errors of the measurements can be evaluated (e.g. using Equations 3.30a, 3.30b, 3.33a, 3.33b, 3.34a and 3.334b) and, consequently, the preferred gauge length of the sensor can be determined.

In cases where the distribution of strain is not known, the assumption of a parabolic distribution can be taken with good accuracy if the loads are linearly distributed, and linear or broken-line distributions can be assumed for the concentrated loads.

While designing the sensing network on a structure (the position and the gauge length of the sensors) it is important to consider all the particularity points that can be identified from a structural analysis and to apply the principles presented in this section, taking into account the aims of monitoring.

3.2.5 Sensor Measurement in Inhomogeneous Materials: Minimal Gauge Length

Reinforced concrete structures, although truly inhomogeneous (at the meso-level), are considered as homogeneous (at the macro-level) during design and structural analysis. The idea of using long-gauge sensors follows the philosophy of reinforced concrete: the structural condition of inhomogeneous material is assessed considering it as being homogeneous.

In an ideally homogeneous material, without local defects or discontinuities (cracks), the elastic strain field follows the theoretical models and is described by the equations presented in Section 3.1.2. This, along with the considerations developed in Sections 3.2.2 and 3.2.3, leads to the conclusion that, in an ideally homogeneous material, the use of a short gauge length is not an issue. In Section 3.2.4 it was even demonstrated that a shorter gauge length can offer better accuracy.

In contrast, inhomogeneous materials, and notably materials with discontinuities (e.g. reinforced concrete), cannot be monitored with short-gauge sensors. Monitoring of inhomogeneous materials at the structural level can only be performed using long-gauge sensors. In this section, the minimal gauge length of such a sensor will be analysed.

The theory presented in this section is demonstrated on a simply tensioned reinforced concrete beam as presented in Figure 3.15. Let three sensors with equal gauge length be installed parallel to the elastic line of the beam as follows: two sensors have anchoring points belonging to the same cross-sections (sensor 1 and sensor 2) and the third sensor (sensor 3) is shifted slightly along the axis; that is, its anchor pieces do not belong to the same cross-sections as the anchor pieces of other two sensors, as shown in Figure 3.15 (top).

Owing to tension, the reinforced concrete element will crack; and let us suppose that the first crack occurs exactly in one of the cross-sections where sensor 1 and sensor 2 have their anchor pieces. In the worst case, the anchor pieces will belong to two separate blocks of concrete (anchor piece of sensor 1 belongs to block 1 and anchor piece of sensor 2 belongs to block 0), as shown in Figure 3.15 (bottom). Sensor 3 is only slightly axially shifted; thus, its anchor piece belongs to block 1 and all the three anchor pieces at the other extremity of the beam belong to the same block (block $n + 1$; see Figure 3.15 (bottom)).

Since the strain field in the beam is constant all three sensors are supposed to measure the same value, which is equal to the average strain in the beam. However, due to cracking and

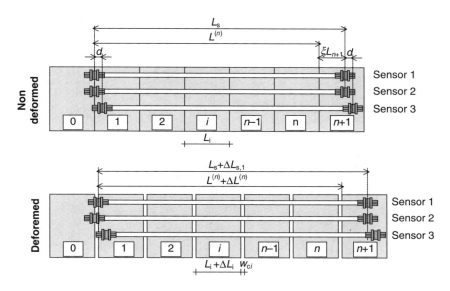

Figure 3.15   Three sensors installed at nearly the same position before cracking of concrete (top) and their worst-case position after the cracking of concrete. L_s is the gauge length of sensors; L_i is the length of concrete block i; $L^{(k)}$ is the sum of lengths of the first k concrete blocks ($L^{(k)} = L_1 + L_2 + \ldots + L_k$); ξ is the factor showing how deep sensors 1 and 2 enter in to the block $n + 1$, $0 < \xi \leq 1$; d is the axial shift of sensor 3 with respect to sensors 1 and 2, $d < L_1$, $d < L_2$; $\Delta L_{s,j}$ is the change in length (deformation) of sensor j, $j = 1, 2, 3$; ΔL_i is the change in length (deformation) of concrete block i; $\Delta L^{(k)}$ is the sum of length changes of the first k concrete blocks ($\Delta L^{(k)} = \Delta L_1 + \Delta L_2 + \ldots + \Delta L_k$); $w_{c,i}$ is the opening of the crack between the blocks i and $i + 1$.

imperfection in the sensors' positions, the measurements of sensors 1, 2 and 3 will be different from each other and from the calculated average strain of the beam. This error is analysed below.

In order to simplify the analysis let us assume that each block in Figure 3.15 has approximately the same length, approximately the same deformation and let us assume that each crack opening has approximately the same value; that is, let

$$L_i \approx L_c = \frac{L^{(n)}}{n}, i = \overline{1, n+1}; \quad \Delta L_i \approx \Delta L_c, i = \overline{1, n+1}; \quad w_{c,i} \approx w_c, i = \overline{0, n+1}$$

$$(3.39)$$

where L_c is the approximate length of each block, ΔL_c is the approximate deformation of each block and w_c is the approximate value of each crack opening.

The approximate average strain in the beam is given by the following expression:

$$\varepsilon_{eq,s} = \varepsilon^{(n)} = \frac{\Delta L^{(n)}}{L^{(n)}} = \frac{\sum_{i=1}^{n} \Delta L_i + \sum_{i=1}^{n} w_{c,i}}{\sum_{i=1}^{n} L_i} \approx \frac{\Delta L_c + w_c}{L_c} \qquad (3.40)$$

The approximate average strains measured by sensor 1, sensor 2 and sensor 3 are

$$
\varepsilon_{s,1} = \frac{\Delta L_{s,1}}{L_s} = \frac{\displaystyle\sum_{i=1}^{n} \Delta L_i + \Delta(\xi L_{n+1}) + \sum_{i=1}^{n} w_{c,i}}{\displaystyle\sum_{i=1}^{n} L_i + \xi L_{n+1}} \approx \frac{\Delta L_c + w_c}{L_c} - \frac{\xi w_c}{(n+\xi)L_c} \tag{3.41a}
$$

$$
\varepsilon_{s,2} = \frac{\Delta L_{s,2}}{L_s} = \frac{\displaystyle\sum_{i=1}^{n} \Delta L_i + \Delta(\xi L_{n+1}) + w_{c,0} + \sum_{i=1}^{n} w_{c,i}}{\displaystyle\sum_{i=1}^{n} L_i + \xi L_{n+1}} \approx \frac{\Delta L_c + w_c}{L_c} + \frac{(1-\xi)w_c}{(n+\xi)L_c}
$$
$$
\tag{3.41b}
$$

$$
\varepsilon_{s,3} = \frac{\Delta L_{s,3}}{L_s} = \frac{\Delta(L_1 - d) + \displaystyle\sum_{i=2}^{n} \Delta L_i + \Delta(\xi L_{n+1} + d) + \sum_{i=1}^{n} w_{c,i}}{\displaystyle\sum_{i=1}^{n} L_i + \xi L_{n+1}} \approx \varepsilon_{s,1} \tag{3.41c}
$$

respectively. The approximate absolute error of the sensors are

$$
\delta\varepsilon_{s,3} \approx \delta\varepsilon_{s,1} = \varepsilon_{s,1} - \varepsilon_{eq,s} \approx -\frac{\xi w_c}{(n+\xi)L_c} = -\xi\frac{w_c}{L_s} \tag{3.42a}
$$

$$
\delta\varepsilon_{s,2} = \varepsilon_{s,2} - \varepsilon_{eq,s} \approx \frac{(1-\xi)w_c}{(n+\xi)L_c} = (1-\xi)\frac{w_c}{L_s} \tag{3.42b}
$$

The parameter ξ varies in a range 0 to 1; thus, the approximate absolute error of the average strain measured with a sensor ranges between the values $-w_c/L_s$ and w_c/L_s:

$$
-\frac{w_c}{L_s} \le \delta\varepsilon_s \le \frac{w_c}{L_s} \Rightarrow \delta\varepsilon_s \approx \pm\frac{w_c}{L_s} \tag{3.43a}
$$

The estimation of the relative error is obtained by a division of Equation 3.43a with Equation 3.40:

$$
\delta^*\varepsilon_s \approx \pm\frac{w_c}{\Delta L_c + w_c}\frac{L_c}{L_s} \tag{3.43b}
$$

Equations 3.43a and 3.43b demonstrate that a longer sensor gauge length provides results with lower errors and, consequently, with better measurement accuracy. Taking into account the conclusion arrived at from Equations 3.43a and 3.43b and the considerations from Section 3.2.3, one can conclude that, in the case of inhomogeneous material, short-gauge sensors prevail for material monitoring and long-gauge sensors are more suitable for structural monitoring.

3.2.6 General Principle in the Determination of Sensor Gauge Length

The conclusions arrived at in Sections 3.2.4 and 3.2.5 can be in contradiction: long-gauge sensors are needed for inhomogeneous materials, but, in the case of parabolic or broken-line strain distribution, shorter sensors would provide results with lower errors. Consequently, in inhomogeneous materials with parabolic or broken-line strain distribution it is practically impossible to perform strain monitoring without an error generated by sensor gauge length.

In homogeneous materials, short-gauge sensors can be used and, consequently, the errors due to sensor gauge length can be better controlled. However, the use of long-gauge sensors is still recommended because they cover bigger areas in the structure and, therefore, the probability of detecting a malfunction, damage or critical strain is increased. Thus, for homogeneous materials also, monitoring at the structural level is better performed using long-gauge sensors.

The general principles for selection of an appropriate sensor gauge length depend on the application type and construction material and are presented in Figure 3.16.

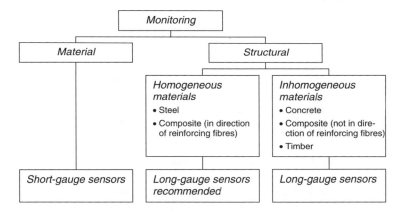

Figure 3.16 General principle of selection of appropriate sensor gauge length depending on application and construction material.

Determination of the gauge length of sensors is to be performed using the equations presented in Sections 3.2.4 and 3.2.5, in a manner so as to minimize errors.

3.2.7 Distributed Strain Sensor Measurement

Distributed fibre-optic sensing presents unique features that have no match in conventional sensing techniques. The ability to measure the strain at thousands of points along a single fibre is particularly interesting for the monitoring of large structures, such as bridges, pipelines, flow lines, oil wells, dams and dykes.

Although distributed sensors are sensitive to strain at every point along the sensing optical fibre, they measure at discrete points that are spaced by a constant value, called the sampling interval, and the measured parameter is actually an average strain measured over a certain length, called the spatial resolution. The explanation of distributed sensor measurement is presented in Figure 3.17.

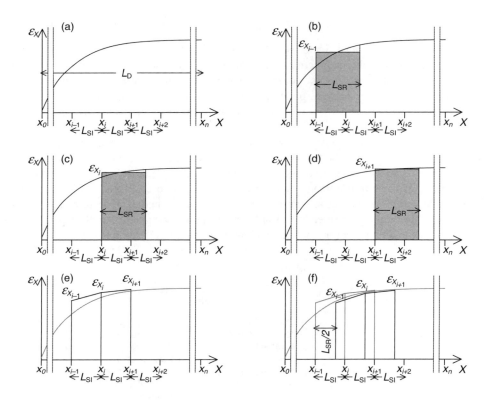

Figure 3.17   Explanation of distributed sensor measurement.

Let us consider a distributed sensor with a total length L_D measured with a sampling interval of L_{SI} and a spatial resolution L_{SR}, as presented in Figure 3.17a.

Let x be the coordinate along the distributed sensor, $n = \text{integer}(L_D/L_{SI})$, x_0 is the coordinate of the first point on the sensor and $x_i = x_0 + iL_{SI}$, $i = 1, 2, 3, \ldots, n$, are coordinates of the points defined by the sampling interval; see Figure 3.17a.

For each point with coordinate x_i the strain is averaged over the segment $[x_i, x_i + L_{SR}]$ as presented in Figure 3.17b–d, and the value of the measurement is attributed to the point x_i as presented in Figure 3.17e.

Finally, the strain diagram of the measurement obtained is shifted horizontally by $L_{SR}/2$, as presented in Figure 3.17f, in order to attribute the average strain measurements to the middle point of the averaging segment (similar to as presented in Section 3.2.4).

The above discussion supports the statement that distributed measurement performed using a distributed sensor with length L_D and sampling interval L_{SI} provides for the same information as n discrete (point) sensors ($n = \text{integer}(L_D/L_{SI})$) with a gauge length of L_{SR}.

Since the distributed sensor provides for the same information as the discrete (point) sensors, all the considerations related to notion of gauge length developed in previous sections are applicable to the notion of spatial resolution of distributed sensors.

This general principle, however, is not valid for abrupt strain changes or concentrated strains (such as generated by cracks). In this case the measurement resulting from a distributed sensing system can be unpredictable and possibly lead to measurement errors. Advanced algorithms have been developed to deal with such situations. Appropriate sensor designs, allowing for controlled strain redistribution over a length compatible with the spatial distribution, can minimize these problems (Glišić *et al.,* 2007).

3.3 Interpretation of strain measurement

3.3.1 Introduction

After the measurement is performed it is important to interpret it correctly. The process of interpretation of measurements consists of two steps:

1. It is important to verify that the measurement is reliable (i.e. correctly performed from the technical point of view and free of handling and sensor malfunctioning errors).
2. After the reliability of measurement is confirmed it is important to identify and evaluate strain components present in the measurement (elastic, plastic, creep, shrinkage, thermal strain and so on) and to detect potential structural malfunctions or damage.

Although apparently uncomplicated, a correct interpretation of strain measurement is often not simple. In order to verify the reliability of the measurement it is necessary to have expertise in the monitoring system employed. Identification of strain components depend on loads and the construction material's properties, but their evaluation depends of several influences, and the errors and uncertainties related to these influences:

- Numerical models used to estimate the strain evolution
- Actual strain sources, resulting strain components and their magnitudes
- Constitutive laws of materials and related constants (mechanical and physical properties of material, such as elasticity, linearity, homogeneity, isotropy, rheologic behaviour, thermal capacity and conductivity)
- Actual geometric properties of the structure
- Actual sensor position in the structure
- Sensor properties (specifications), such as gauge length, accuracy and long-term stability
- Other.

The influences listed above might not be mutually independent, and it is recommended to analyse them simultaneously rather than separately.

Previously, the different sources of strain in different materials were presented in detail in Section 3.1 and the influence of gauge length on measurement was presented in Sections 3.2.4, 3.2.5 and 3.2.7. Below, we highlight the sources of errors and illustrate the interpretation of measurement with an on-site example.

3.3.2 Sources of Errors and Detection of Anomalous Structural Condition

Two main sources or errors are related to uncertainties (e.g. whether loads are present on a structure or not at the moment of measurement) and the errors generated while estimating different quantitative values necessary to interpret measurement (e.g. to determine the elastic strain it is first necessary to determine the thermal strain component, which depends on estimation of the thermal expansion coefficient of the construction material; the error in its estimation will generate error in interpretation of the measurement).

Based on the considerations presented in Section 3.1, a general numerical model describing the strain field in planar linear structural elements (beams with constant cross-section loaded only in one plane) with respect to the axis of the element (axis z; see Figure 3.4) and built from linear viscoelasto-plastic material is proposed as follows:

$$\varepsilon_{est,tot,t}(y, z) = \frac{\sigma_{E,t}(y, z)}{E_t}(1 + K_{\varphi,equ} f_{\varphi,equ,t}) + \varepsilon_{P,t}(y, z) + \alpha_T \Delta T_t(y, z) + \varepsilon_{sh,final} f_{sh,t}$$

$$(3.44)$$

where $K_{\varphi,equ}$ is the equivalent creep coefficient (for more details, see Section 3.3.3) and $f_{\varphi,equ,t}$ is the equivalent creep function at time t (for more details, see Section 3.3.3)

$$\sigma_{E,t}(y, z) = \frac{N_t(z)}{A_t} + \frac{M_t(z)}{I_t} y_t = E_t \varepsilon_{E,t}(y, z) - \text{Stress at time } t$$

The other notions are presented in Section 3.1.

The numerical model provides for estimated total strain and estimated strain components. The error of estimated total strain at a point with coordinates y and z is equal to the difference between the estimated and real strain values, as presented in Equation 3.45:

$$\delta\varepsilon_{est,tot,t}(y, z) = \varepsilon_{est,tot,t}(y, z) - \varepsilon_{real,tot,t}(y, z) \qquad (3.45)$$

The error of the model depends on errors in evaluation of all the parameters included in the model. Consequently, in Equation 3.44, the error in estimation of the total strain depends on:

1. Appropriateness of the constitutive material laws employed; more precisely:
   a. Stress–strain relation in elastic domain
   b. Stress–strain relation in plastic domain
   c. Thermal strain model
   d. Rheologic strain models
   e. Presence of homogeneity discontinuities (see Section 3.2.3)
   f. Other.
2. Error in determination of internal forces, normal (axial) force $N_t(z)$ and bending moment $M_t(z)$ depends on:
   a. External loads
   b. Static system of the structure
   c. Stiffness of joints between structural members
   d. Other.
3. Error in determination of geometrical properties, cross-sectional area A_t, moment of inertia I_t and distance from centre of gravity y_t depends on:
   a. Quality of construction (tolerance) of structural members
   b. Structural changes during the service (e.g. structural cracks in concrete)

c. Degradation due to external influences (e.g. reduction of cross-section due to corrosion in steel)

d. Other.

4. Error in determination of basic mechanical and thermal properties of construction material, such as:

   a. Young's modulus E_t

   b. Thermal expansion coefficient $\alpha_{T,t}$.

5. Error in determination of rheologic parameters:

   a. Creep constant and function

   b. Shrinkage final value and shrinkage function.

6. Other.

It is important to highlight that, in general, all the parameters discussed can change in time. For example, the normal force and bending moment depend on the applied load, which can vary in time. Also, the Young's modulus of concrete changes in time, as do the geometric properties of the cracked concrete.

For inhomogeneous materials, Equation 3.45 can give big errors since the model presented in Equation 3.44 represents a continuous function and the real strain distribution can be discontinuous due to cracks or inclusions (see Section 3.2.3). That is why the error of the numerical model at a point with coordinates y and z is better estimated using Equation 3.46, which is based on strain averaged over the length L:

$$\delta\varepsilon_{\text{est,tot},L,t}(y, z) = \frac{1}{L} \int_{z-(L/2)}^{z+(L/2)} \varepsilon_{\text{est,tot},t}(y, \zeta)\,d\zeta$$

$$-\frac{1}{L}\left(\int_{z-(L/2)}^{z+(L/2)} \varepsilon_{\text{real,tot},t}(y, \zeta)\,d\zeta + \sum_{i=n_{c,z-(L/2)}}^{n_{c,z+(L/2)}} \Delta w_{\text{real,c},i,t}(y) \right.$$

$$\left. + \sum_{j=n_{\text{incl},z-(L/2)}}^{n_{\text{incl},z+(L/2)}} \Delta w_{\text{real,incl},j,t}(y) \right)$$

$$(3.46)$$

where the first term on the right-hand side of Equation 3.46 represents estimated averaged strain and the second term represents averaged real strain over the length L, namely between the points A and B with coordinates $A(y, z - (L/2))$ and $B(y, z + (L/2))$ respectively.

The error of the model depends on all the errors described in the above points (1)–(6). For different errors in estimation of the parameters listed in points (1)–(6) and taking into account Equation 3.44, a different model error is obtained (Equation 3.46). In order to simplify the error analysis, the notion of maximum absolute error of the model is introduced, which is defined as

$$\delta\varepsilon_{\text{est,tot},L,t}^{\max}(y, z) = \max\{|\delta\varepsilon_{\text{est,tot},L,t}(y, z)|\} \qquad (3.47)$$

where the maximum is performed over the absolute value of the model error for all the possible errors generated in determination of the parameters listed in points (1)–(6) and taking into account Equation 3.44 (e.g. δE, δA, δI, etc.). Supposing that the model ideally represents the real strain and that the errors in determination of all the parameters involved are known, then it is possible to determine the maximal absolute error of model using error calculus.

When the maximum absolute error is correctly calculated, the value of the estimated strain at a point with coordinates (y, z) is often expressed as $\varepsilon_{est,tot,L,t}(y, z) \pm \delta\varepsilon_{est,tot,L,t}^{max}(y, z)$ and the real strain value at the same point fulfils the following condition:

$$\varepsilon_{est,tot,L,t}(y, z) - \delta\varepsilon_{est,tot,L,t}^{max}(y, z) \leq \varepsilon_{real,tot,L,t}(y, z) \leq \varepsilon_{est,tot,L,t}(y, z) + \delta\varepsilon_{est,tot,L,t}^{max}(y, z)$$

$$(3.48)$$

The maximal relative error of the estimated total strain (of the model) is defined as

$$\delta^{*}\varepsilon_{est,tot,L,t}^{max}(y, z) = \left| \frac{\delta\varepsilon_{est,tot,L,t}^{max}(y, z)}{\varepsilon_{real,tot,t}(y, z)} \right|$$

$$(3.49)$$

and it is commonly expressed as a percentage (Equation 3.49 is multiplied by 100). When the maximal relative error is correctly calculated at a point with coordinates (y, z), the real strain value at the same point fulfils the following condition:

$$\frac{\left| \varepsilon_{est,tot,L,t}(y, z) \right|}{1 + \delta\varepsilon_{est,tot,L,t}^{max}(y, z)} \leq \left| \varepsilon_{real,tot,L,t}(y, z) \right| \leq \frac{\left| \varepsilon_{est,tot,L,t}(y, z) \right|}{1 - \delta\varepsilon_{est,tot,L,t}^{max}(y, z)}$$

$$(3.50)$$

Since the sensor provides for measurement of average total strain, the error of measurement is defined as the difference between the measured total average strain value and the real total average strain value. The error of measurement of a sensor with gauge length L_s and anchoring points $A(y_s, z_s - (L_s/2))$ and $B(y_s, z_s + (L_s/2))$ is given by

$$\delta\varepsilon_{m,t}(y_s, z_s) = \varepsilon_{m,t}(y_s, z_s)$$

$$- \frac{1}{L_s} \left(\int_{z_s-(L_s/2)}^{z_s+(L_s/2)} \varepsilon_{real,tot,t}(y_s, \zeta) \, d\zeta + \sum_{i=n_{c,z_s-(L_s/2)}}^{n_{c,z_s+(L_s/2)}} \Delta w_{real,c,i,t}(y_s) \right.$$

$$\left. + \sum_{i=n_{incl,z_s-(L_s/2)}}^{n_{incl,z_s+(L_s/2)}} \Delta w_{real,incl,i,t}(y_s) \right)$$

$$(3.51)$$

where (y_s, z_s) represents the coordinates of the sensor's middle point C–C(y_s, z_s).

The measurement error depends on the following parameters:

1. The sensor's actual position in the structure, coordinates of points A and B and, consequently:
   a. Coordinate of the sensor's middle point C
   b. Parallelism of sensor to the elastic line of the beam (axis z).

2. The sensor specifications
   a. Accuracy of the sensor and the monitoring system
   b. Gauge length of the sensor (see Sections 3.2.4 and 3.2.5).
3. Other.

Similar to the maximal absolute error of estimated strain (Equation 3.47), the maximal absolute error of measurement is defined by Equation 3.52. The error of measurement depends on all the errors described in the above points (1)–(3). For different errors in estimation of parameters listed in points (1)–(3) and taking into account Equation 3.51, a different measurement error is obtained (Equation 3.51). In order to simplify the error analysis, the notion of maximal absolute error of the measurement is introduced, which is defined as

$$\delta\varepsilon_{m,t}^{max} = \max\{|\delta\varepsilon_{m,t}(y_s, z_s)|\} \tag{3.52}$$

where the maximum is performed over the absolute value of the measurement error for all the possible errors generated in determination of the parameters listed in points (1)–(3) above and taking into Equation 3.44 (e.g. δL_s, accuracy of the sensor $\delta\varepsilon_{sen}$, etc.). If we suppose that the errors in determination of all the parameters involved are known, then it is possible to determine the maximal absolute error of measurement using error calculus.

Once the maximal absolute error is correctly calculated, the value of the measured strain at a point with coordinates (y_s, z_s) is often expressed as $\varepsilon_{m,t}(y_s, z_s) \pm \delta\varepsilon_{m,t}^{max}(y_s, z_s)$ and the real strain value at the same point fulfils the following condition:

$$\varepsilon_{m,t}(y_s, z_s) - \delta\varepsilon_{m,t}^{max}(y_s, z_s) \le \varepsilon_{real,tot,L,t}(y_s, z_s) \le \varepsilon_{m,t}(y_s, z_s) + \delta\varepsilon_{m,t}^{max}(y_s, z_s) \tag{3.53}$$

The maximal relative error of the measured total strain is defined as

$$\delta^*\varepsilon_{m,t}^{max}(y_s, z_s) = \left| \frac{\delta\varepsilon_{m,t}^{max}(y_s, z_s)}{\varepsilon_{real,tot,t}(y_s, z_s)} \right| \tag{3.54}$$

and it is commonly expressed as a percentage (Equation 3.54 is multiplied by 100). Once the maximal relative error is correctly calculated at a point with coordinates (y_s, z_s), the real strain value at the same point fulfils the following condition:

$$\frac{|\varepsilon_{m,t}(y_s, z_s)|}{1 + \delta\varepsilon_{m,t}^{max}(y_s, z_s)} \le |\varepsilon_{real,tot,L,t}(y, z)| \le \frac{|\varepsilon_{m,t}(y_s, z_s)|}{1 - \delta\varepsilon_{m,t}^{max}(y_s, z_s)} \tag{3.55}$$

The real total strain is measured with the maximal absolute error presented in Equation 3.52 and is estimated using the general model presented in Equation 3.44 with maximal absolute error presented in Equation 3.47. The maximal allowed difference between the measurement and estimated total strain is given by

$$\delta\varepsilon_{m-est,t}^{max}(y_s, z_s) = \delta\varepsilon_{m,t}^{max}(y_s, z_s) + \delta\varepsilon_{est,tot,L_s,t}^{max}(y_s, z_s) \tag{3.56}$$

The errors $\delta\varepsilon_{m,t}^{max}(y_s, z_s)$ and $\delta\varepsilon_{est,tot,L_s,t}^{max}(y_s, z_s)$ are mutually independent and it is not likely that both measured and estimated strains are simultaneously at the limits of errors. Consequently, the typical maximal difference between the measurement and estimated total strain can alternatively be calculated using

$$\delta\varepsilon_{m-est,t}^{typ} = \sqrt{\delta\varepsilon_{m,t}^2 + \delta\varepsilon_{est,tot,L_s,t}^2} < \delta\varepsilon_{m-est,t}^{max} \tag{3.57}$$

Good agreement between measured and estimated total strain values is a confirmation of a sound structural condition. Supposing that the measurement of the sensor having a middle point with coordinates (y_s, z_s) is reliable, the numerical model of structural behaviour is appropriate, all the sources of error are taken into account and the errors of the numerical model and measurement correctly calculated, then the following condition must be fulfilled (combination of Equations 3.48 and 3.53):

$$\varepsilon_{est,tot,L_s,t}(y_s, z_s) - \delta\varepsilon_{m-est,t}^{max/typ}(y_s, z_s) \leq \varepsilon_{m,t}(y_s, z_s) \leq \varepsilon_{est,tot,L_s,t}(y_s, z_s) + \delta\varepsilon_{m-est,t}^{max/typ}(y_s, z_s)$$

$$\tag{3.58}$$

If the condition presented in Equation 3.58 is not fulfilled, then:

A. There is an anomaly in structural behaviour
B. The errors $\delta\varepsilon_{m,t}^{max}(y_s, z_s)$, $\delta\varepsilon_{est,tot,L_s,t}^{max}(y_s, z_s)$ and, consequently, $\delta\varepsilon_{m-est,t}^{max}(y_s, z_s)$ are not correctly calculated
C. The sensor measurement is not reliable
D. The numerical model is not appropriate
E. Two or more facts presented in points (A)–(D) are valid.

The use of a typical value in Equation 3.67 is more conservative, whereas the use of a maximal allowed value takes into account the possibility that the measured and estimated strains can simultaneously be at the limits of the errors and, therefore, represent the ultimate allowed difference between the estimated and measured strains.

Surpassing the typical maximal difference from time to time can often be tolerated, but systematically exceeding the typical maximal difference as well as surpassing the maximal allowed difference may signal an anomaly in structural behaviour. If that happens then detailed analysis and examination of structural behaviour is advisable.

3.3.3 Determination of Strain Components and Stress from Total Strain Measurement

The different strain components for different materials are presented in Section 3.1. Besides detection of anomalous structural behaviour, it is often of interest to determine the stresses in the structure. Fibre-optic sensors provide for total average strain measurement, and in order to determine the stress from measurement it is necessary to determine the elastic strain component. To simplify the presentation, a unidirectional total strain state is analysed in this section (beams). Similar analysis can be performed for bi- and tri-axial strain states, taking into account the stress–strain relations developed in the *Strength of Materials* (Brčić, 1989) and in the *Theory of Elasticity* (Timoshenko and Goodier, 1970).

In the case of linear-elastic materials, such as composites, only elastic and thermal strain components are present (see Table 3.1); thus, the total strain is given by

$$\varepsilon_{\text{tot},t}(y, z) = \frac{\sigma_{\text{E},t}(y, z)}{E} + \alpha_\text{T} \Delta T_t(y, z) \qquad (3.59)$$

where the first term on the right-hand side of the equation represents the elastic strain and the second term represents thermal strain.

Supposing that the total strain and temperature changes are monitored, the strain components are determined from measurements as follows:

$$\varepsilon_{\text{Tm},t}(y_\text{s}, z_\text{s}) = \alpha_\text{T} \Delta T_{\text{m},t}(y_\text{s}, z_\text{s}) \qquad (3.60)$$

$$\varepsilon_{\text{Em},t}(y_\text{s}, z_\text{s}) = \varepsilon_{\text{m},t}(y_\text{s}, z_\text{s}) - \alpha_\text{T} \Delta T_{\text{m},t}(y_\text{s}, z_\text{s}) \qquad (3.61)$$

where $\Delta T_{\text{m},t}(y_\text{s}, z_\text{s})$ is the measured temperature variation at time t, $\varepsilon_{\text{Tm},t}(y_\text{s}, z_\text{s})$ is the thermal strain at time t determined from temperature monitoring and $\varepsilon_{\text{Em},t}(y_\text{s}, z_\text{s})$ is the elastic strain at time t determined from the total strain ($\varepsilon_{\text{m},t}(y_\text{s}, z_\text{s})$) and temperature monitoring.

The stress determined from monitoring is calculated using

$$\sigma_{\text{Em},t}(y_\text{s}, z_\text{s}) = E(\varepsilon_{\text{m},t}(y_\text{s}, z_\text{s}) - \alpha_\text{T} \Delta T_{\text{m},t}(y_\text{s}, z_\text{s})) \qquad (3.62)$$

Young's modulus E and thermal expansion coefficient α_T are constant for composite materials and are usually either determined in tests or calculated based on data and algorithms found in the literature.

It is important to highlight that stress calculated in Equation 3.62 does not represent absolute value of stress, but the stress change (relative stress) with respect to reference time. Equation 3.62 can be used to determine the absolute stress only if the monitored structure was not stressed at reference time (zero stress at reference time).

In order to simplify the presentation, for all Equations presented in further text of this section (Equation 3.63a to 3.77), it is assumed that the reference time coincides with zero stress and zero strain state of the structure. In this manner measurement at time t practically represent the absolute strain value.

Let $\delta\sigma^{\text{max}}_{\text{Em},t}(y_\text{s}, z_\text{s})$ be the maximal error of the determined strain obtained by applying error calculus principles and Equation 3.62, then a sound structure fulfils the following conditions:

$$\sigma_{\text{Em},t}(y_\text{s}, z_\text{s}) + \delta\sigma^{\text{max}}_{\text{Em},t}(y_\text{s}, z_\text{s}) \leq \frac{\sigma^+_\text{u}}{\beta^+} \qquad (3.63a)$$

$$\frac{\sigma^-_\text{u}}{\beta^-} \leq \sigma_{\text{Em},t}(y_\text{s}, z_\text{s}) - \delta\sigma^{\text{max}}_{\text{Em},t}(y_\text{s}, z_\text{s}) \qquad (3.63b)$$

where $\sigma^\pm_\text{u}$ is the ultimate tensile (superscript '+') and compressive (superscript '−') stress and $\beta^\pm$ are the corresponding safety factors used in structural design ($\beta^\pm \geq 1$).

Similar to the discussion presented in Section 3.3.2, if one of the conditions presented in Equations 3.63a and 3.63b is not fulfilled, then:

a. There is anomaly in structural behaviour, e.g. the structure is overloaded
b. The $\delta\sigma^{\text{max}}_{\text{Em},t}(y_\text{s}, z_\text{s})$ error is not correctly calculated
c. The sensor measurements (strain and temperature) are not reliable

d. The numerical model (Equations 3.59 and 3.62) is not appropriate
e. Two or more facts presented in points (a)–(d) are valid.

For linear elasto-plastic materials, such as metals (steel and aluminium), besides elastic and thermal strains, the plastic strain component can be present (see Table 3.1) and the total strain is given by

$$\varepsilon_{\text{tot},t}(y, z) = \frac{\sigma_{\text{E},t}(y, z)}{E} + \varepsilon_{\text{P},t}(y, z) + \alpha_{\text{T}}\Delta T_t(y, z) \tag{3.64}$$

The thermal strain is determined using Equation 3.60. Plastic strain is commonly not allowed in the service state of the structure; consequently, the verification that plastic strain is not present is the first to be performed, as shown in Equation 3.65 (see also Figure 3.3):

$$\varepsilon_{\text{P}_m,t}(y_s, z_s) = 0 \Leftrightarrow \varepsilon_{\text{S}_m,t}(y_s, z_s) = \varepsilon_{\text{m},t}(y_s, z_s) - \alpha_{\text{T}}\Delta T_{\text{m},t}(y_s, z_s) - \varepsilon_y \leq 0 \tag{3.65}$$

If the condition presented in Equation 3.65 is fulfilled, then the structure is in the elastic state and Equations 3.61 and 3.62 are applicable for calculation of elastic strain and stress; the sound structure fulfils the conditions in Equations 3.63a and 3.63b.

If the condition presented in Equation 3.65 is not fulfilled, then the structure is in the plastic state and its soundness is imperilled. The elastic and plastic strain components and the stress are then calculated as follows (see Figure 3.3 and Equation 3.6):

$$\varepsilon_{\text{E}_m,t}(y_s, z_s) = \varepsilon_y \tag{3.66}$$

$$\varepsilon_{\text{P}_m,t}(y_s, z_s) = \varepsilon_{\text{m},t}(y_s, z_s) - \alpha_{\text{T}}\Delta T_{\text{m},t}(y_s, z_s) - \varepsilon_y \tag{3.67}$$

$$\sigma_{\text{E}_m,t}(y_s, z_s) = E\varepsilon_y = \sigma_u \tag{3.68}$$

The conditions in Equations 3.63a and 3.63b can never be fulfilled for stress calculated in Equation 3.68.

Viscoelasto-plastic materials, such as concrete or timber, can be analysed using simplified the linear viscoelasto-plastic model proposed in Equation 3.44, or using a more accurate nonlinear viscoelasto-plastic model, the development of which exceeds the scope of this book. The main challenge for these materials is determination of rheologic components, creep and shrinkage, which is to be performed using either separate experimental methods or numerical models, as discussed in Sections 3.1.4 and 3.1.5.

A particular issue in determination of the creep component is its dependency on elastic strain and, in the case of concrete, the dependency of the creep coefficient on the concrete maturity at the time of loading. In the other words, it is necessary to know an accurate time history of loading. For example, the same concrete element loaded at different times t_A and t_B ($t_A < t_B$) with different stresses $\sigma_{\text{E,A}}$ and $\sigma_{\text{E,B}}$ ($\sigma_{\text{E,A}} < \sigma_{\text{E,B}}$) has different values of creep coefficients and different values of the creep function and, consequently, different creep evolutions, depending on the load case, as shown in Figure 3.18.

At an arbitrary time of measurement t ($t_A < t_B < t$), the sum of elastic strain and creep can have the same value; but if the time of loading is not known, then it is impossible to determine the

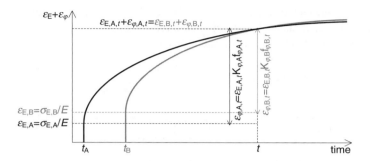

Figure 3.18   The same concrete element can have two different creep evolutions depending on the time of loading and the maturity of concrete at the time of loading.

elastic strain (and the creep) from the measurement (i.e. whether the measurement corresponds to load case 'A' or 'B').

Determination of creep becomes more complex if the load changes significantly in time. The creep in a concrete element loaded in n steps at different times t_i, $i \in \{1, 2, \ldots, n\}$, elapsed from pouring of the concrete can be determined using the following formula:

$$\varepsilon_{\varphi,t} = \sum_{i=1}^{n} \left(\frac{\Delta\sigma_{\mathrm{E},t_i}}{E} K_{\varphi i,t} f_{\varphi i,t} \right) = \left(\sum_{i=1}^{n} \frac{\Delta\sigma_{\mathrm{E},t_i}}{E} \right) K_{\varphi,\mathrm{equ}}^{n} f_{\varphi,\mathrm{equ},t}^{n} = \varepsilon_{\mathrm{E},t} K_{\varphi,\mathrm{equ}}^{n} f_{\varphi,\mathrm{equ},t}^{n}$$

(3.69)

Each step of load $\Delta\sigma_{\mathrm{E},t_i}$ generates a corresponding creep coefficient $K_{\varphi i}$ and creep function $f_{\varphi i,t}$; thus, determination of the equivalent creep coefficient $K_{\varphi,\mathrm{equ}}^{n}$ and equivalent creep function $f_{\varphi,\mathrm{equ},t}^{n}$ at time t ($t_n < t$) is particularly challenging.

Let $\varepsilon_{\mathrm{shm},\mathrm{final}}$ be the final value of shrinkage, $f_{\mathrm{shm},t}$ the shrinkage function, $K_{\varphi\mathrm{m},\mathrm{equ}}$ the equivalent creep coefficient and $f_{\varphi\mathrm{m},\mathrm{equ},t}$ the equivalent creep function determined using methods presented in Section 3.1.4 and 3.1.5. The shrinkage is determined in measurement as follows:

$$\varepsilon_{\mathrm{shm},t} = \varepsilon_{\mathrm{shm},\mathrm{final}} f_{\mathrm{shm},t}$$

(3.70)

Thermal strain is determined using Equation 3.60. The strain generated by loads, including the elastic strain, creep and plastic strain, is determined using

$$\varepsilon_{\mathrm{E}_{\mathrm{m}},t}(y_{\mathrm{s}}, z_{\mathrm{s}})(1 + K_{\varphi\mathrm{m},\mathrm{equ}} f_{\varphi\mathrm{m},\mathrm{equ},t}) + \varepsilon_{\mathrm{P}_{\mathrm{m}},t}(y_{\mathrm{s}}, z_{\mathrm{s}})$$
$$= \varepsilon_{\mathrm{m},t}(y_{\mathrm{s}}, z_{\mathrm{s}}) - \alpha_{\mathrm{T}}\Delta T_{\mathrm{m},t}(y_{\mathrm{s}}, z_{\mathrm{s}}) - \varepsilon_{\mathrm{shm},\mathrm{final}} f_{\mathrm{shm},t}$$

(3.71)

The left-hand side of Equation 3.71 contains two unknowns: the elastic strain and the plastic strain. The presence of plastic strain is discriminated by the following condition:

$$\varepsilon_{\mathrm{P}_{\mathrm{m}},t}(y_{\mathrm{s}}, z_{\mathrm{s}}) = 0 \Leftrightarrow \varepsilon_{\mathrm{m},t}(y_{\mathrm{s}}, z_{\mathrm{s}}) - \alpha_{\mathrm{T}}\Delta T_{\mathrm{m},t}(y_{\mathrm{s}}, z_{\mathrm{s}})$$
$$- \varepsilon_{\mathrm{shm},\mathrm{final}} f_{\mathrm{shm},t} - \varepsilon_{y}(1 + K_{\varphi\mathrm{m},\mathrm{equ}} f_{\varphi\mathrm{m},\mathrm{equ},t}) \leq 0$$

(3.72)

If the condition presented in Equation 3.72 is fulfilled, then the structure is in the elastic state and Equations 3.73 and 3.74 are applicable for calculation of elastic strain and stress; a sound structure fulfils the conditions in Equations 3.63a and 3.63b.

$$\varepsilon_{\mathrm{E_m},t}(y_\mathrm{s}, z_\mathrm{s}) = \frac{\varepsilon_{\mathrm{m},t}(y_\mathrm{s}, z_\mathrm{s}) - \alpha_\mathrm{T} \Delta T_{\mathrm{m},t}(y_\mathrm{s}, z_\mathrm{s}) - \varepsilon_{\mathrm{sh_m},\mathrm{final}} f_{\mathrm{sh_m},t}}{1 + K_{\varphi_\mathrm{m},\mathrm{equ}} f_{\varphi_\mathrm{m},\mathrm{equ},t}} \tag{3.73}$$

$$\sigma_{\mathrm{E_m},t}(y_\mathrm{s}, z_\mathrm{s}) = E \frac{\varepsilon_{\mathrm{m},t}(y_\mathrm{s}, z_\mathrm{s}) - \alpha_\mathrm{T} \Delta T_{\mathrm{m},t}(y_\mathrm{s}, z_\mathrm{s}) - \varepsilon_{\mathrm{sh_m},\mathrm{final}} f_{\mathrm{sh_m},t}}{1 + K_{\varphi_\mathrm{m},\mathrm{equ}} f_{\varphi_\mathrm{m},\mathrm{equ},t}} \tag{3.74}$$

If the condition presented in Equation 3.72 is not fulfilled, then the structure is in the plastic state and its soundness is imperilled. The elastic and plastic strain components and the stress are then calculated using Equations 3.66, 3.75 and 3.68 respectively (see also Figure 3.3 and Equation 3.6).

$$\varepsilon_{\mathrm{P_m},t}(y_\mathrm{s}, z_\mathrm{s}) = \varepsilon_{\mathrm{m},t}(y_\mathrm{s}, z_\mathrm{s}) - \alpha_\mathrm{T} \Delta T_{\mathrm{m},t}(y_\mathrm{s}, z_\mathrm{s}) - \varepsilon_{\mathrm{sh_m},\mathrm{final}} f_{\mathrm{sh_m},t} - \varepsilon_\mathrm{y}(1 + K_{\varphi_\mathrm{m},\mathrm{equ}} f_{\varphi_\mathrm{m},\mathrm{equ},t})$$

$$\tag{3.75}$$

Similarly, as discussed in the case of linear elasto-plastic material, the conditions in Equations 3.63a and 3.63b can never be fulfilled if a linear viscoelasto-plastic material is in the plastic state.

Discrimination of the plastic state is performed using Equation 3.72, which contains the constants and the functions related to rheologic strain components (creep and shrinkage) that are very challenging to determine. Consequently, discrimination of the plastic state is subject to errors generated by determination of the creep and shrinkage constants and the functions and plastic strain can easily be confused with creep and shrinkage. It is recommended, therefore, to pay particular attention to the presented issues while interpreting the measurement performed on linear viscoelasto-plastic materials.

The nonlinear viscoelasto-plastic model for a material such as concrete is

$$\varepsilon_{\mathrm{est,tot},t}(y, z) = \varepsilon_{\mathrm{E},t}(1 + K_{\varphi,\mathrm{equ}} f_{\varphi,\mathrm{equ},t}) + \varepsilon_{\mathrm{P},t}(y, z) + \alpha_\mathrm{T} \Delta T_t(y, z) + \varepsilon_{\mathrm{sh},\mathrm{final}} f_{\mathrm{sh},t} \tag{3.76}$$

The short-term stress–strain relation in the elastic state is presented schematically in Figure 3.3b and by the following formula:

$$\sigma_{\mathrm{E},t}(y, z) = F_\sigma(\varepsilon_{\mathrm{E},t}(y, z)) \tag{3.77}$$

where F_σ is function describing non-linear short-term stress-strain relation.

The considerations related to the linear viscoelasto-plastic model are applicable in the non-linear case, and Equations 3.69–3.76 are valid for a nonlinear viscoelasto-plastic model with the condition that the linear short-term stress–strain relation ($\sigma_{\mathrm{E},t}(y, z) = E \cdot \varepsilon_{\mathrm{E},t}(y, z)$) is substituted with Equation 3.77.

The above analysis is carried out for absolute strain measurements (zero stress and zero strain state of structure at reference time). If the absolute measurements are not available (sensors installed on already stressed and strained structure), then the actual value of strain for each component (structural, thermal, creep, shrinkage) at reference time must be added to corresponding strain in Equations 3.63a to 3.77).

Table 3.2   Typical values of parameters for different construction materials

Parameter	Concrete	Steel	Composite
$\varepsilon_u^+\,(\mu\varepsilon)$	50 to 100	5000 to 10 000	1500 to 3000
$\varepsilon_u^-\,(\mu\varepsilon)$	-3500	-5000 to $-10\,000$	-1500 to -3000
$\varepsilon_y^+\,(\mu\varepsilon)$	50 to 100	2000	—
$\varepsilon_y^-\,(\mu\varepsilon)$	-1350 to -2000	-2000	—
$K_\varphi\,(-)$	1 to 3 (5.6)	—	—
$\varepsilon_{\mathrm{sh,final}}\,(\mu\varepsilon)$	-200 to -530	—	—
$\alpha_\mathrm{T}\,(\mu\varepsilon\,{}^\circ\mathrm{C}^{-1})$	8 to 12	10 to 12	
$E\,(\mathrm{GPa})$	20 to 50	210	

The determination of actual strain values at reference time may be complex and difficult, and may require destructive methods or advanced modeling. To avoid these issue it is recommended to start monitoring with the birth of the structure, i.e. at time when the structure is not strained nor stressed.

Some typical values of parameters related to interpretation of strain measurements for various materials are given in Table 3.2.

3.3.4 Example of Strain Measurement Interpretation

Recent development in fibre-optic sensors and informatics technology have made SHM of civil structures cost effective. In this example, a part of a large-scale lifetime buildings monitoring programme is presented. The programme started with a pilot project in 2001 in Singapore and since then a significant number of high-rise buildings have been monitored (Glišić *et al.*, 2005). The monitoring aims of this unique programme have been to increase safety, verify performance, control quality, increase knowledge, optimize maintenance costs and evaluate the condition of the structure after an earthquake, impact or terrorist act.

Long-gauge fibre-optic sensors were embedded in the ground-level columns during the construction; thus, monitoring started with the birth of the structure. Based on the results it was possible to evaluate and follow the performance of the buildings over the long term, during every stage of their life. In this section, only the results obtained on one column are presented in order to illustrate the measurement interpretation. A more exhaustive presentation of the complete project is given in Chapters 4 and 5.

The ground columns have been selected as being the most critical elements in the building, whereas the number of sensors was adapted to the available budget. The dominant load in each column is a compressive normal force; therefore, it is supposed that the influence of bending on deformation can be neglected. Consequently, a single sensor per column, installed parallel to the column axis, and not necessary in the centre of gravity of the cross-section, is estimated as sufficient for monitoring at the local column level. The position of the sensor in the column is presented schematically in Figure 3.19a. The length of the sensors is determined with respect to the available height of the column (3.5 m) and on-site conditions; hence, 2 m long sensors were used.

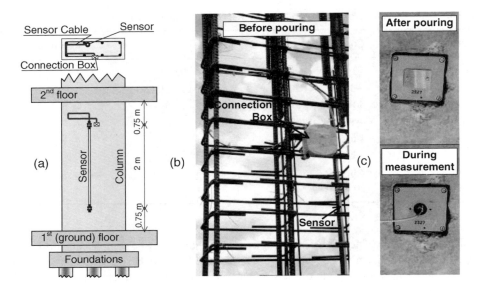

Figure 3.19   (a) Sensor position in column; (b) sensor and connection box installed on a rebar cage; (c) connection box after the pouring of column (courtesy of SMARTEC and Sofotec).

In each column, the sensor was attached on rebars before the pouring of concrete, as shown in Figure 3.19b. The sensor connector was protected with a small connection box, which was also embedded in the concrete (see Figure 3.19). In this way neither the sensor cable nor the connector egresses from the column. The connection box is provided with a small opening to allow access to the sensor connector after the column is poured. Closure of the opening and the connected sensor are shown in Figure 3.19c.

At the time of writing, measurements performed over more than 5 years were collected. To decrease the costs of monitoring, only periodic readings were performed: one session over all the sensors after a new storey was completed, and once every few months after the construction. This periodic manner of collecting data is justified by the fact that no significant issue was detected during the construction phase or later. The very early age measurements are estimated as not important in this project and, therefore, were not performed.

The most important parameters that influenced the performance of monitoring were the schedule of measurements, temperature, loads and the numerical model for data analysis.

The initial measurements were performed after the second storey was completed (recall that the second storey is the first storey built on columns equipped with sensors; see Figure 3.19). This measurement is a reference for all further measurements. The initial measurements before the second storey was built and the measurements after the third storey was built were not registered since the sensors were inaccessible. These missing measurements involved some imperfection in the calculation during data analysis.

The temperature in Singapore ranges between 23 and 33 °C during the day or night independently of the season. This fact, along with the limited budget for monitoring, led to the decision not to monitor the temperature.

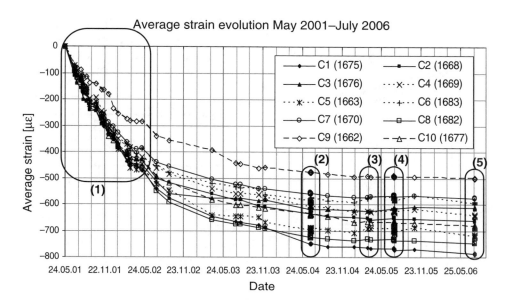

Figure 3.20 Evolution of total average strain in columns monitored over more than 5 years; highlighted periods: (1) construction of 19 storeys; (2) the first 48 h continuous monitoring session performed in July 2004; (3) before and after tremor monitoring; (4) the second 48 h continuous monitoring session performed in July 2005; (5) the third 48 h continuous monitoring session performed in July 2006 (data provided by HDB and SMARTEC).

The diagram presented in Figure 3.20 shows the time-dependent evolution of the average strain in columns monitored during more than 5 years. The following five particularly important periods are highlighted in Figure 3.20: (1) construction of 19 storeys, (2) the first 48 h continuous monitoring session performed in July 2004, (3) before and after tremor monitoring and (4) the second 48 h continuous monitoring session performed in July 2005 and (5) the third 48 h continuous monitoring session performed in July 2006.

In order to illustrate the interpretation of the measurement and to apply the algorithms presented in previous sections, column C1 is analysed in more detail. The concrete was considered to be a linear viscoelasto-plastic material. Based on concrete design, the Young's modulus was considered to be 28 GPa. The creep and shrinkage parameters were calculated using the CEB-FIP, 1990 Model Code, and the load–time history was determined using the design values of load for each storey and the actual time from completion of the individual storeys.

A particular problem in the analysis is the estimation of the errors. The temperature was not measured, but taking into account that the maximal temperature variations in Singapore are 10 °C and using a typical thermal expansion coefficient of concrete (estimated) of 12 $\mu\varepsilon$ °C^{-1}, the maximal absolute error due to lack of temperature monitoring is estimated to be ± 120 $\mu\varepsilon$. However, the 48 h continuous measurements showed maximal variations of 10 $\mu\varepsilon$ for 3 °C; thus, it is more realistic to set the error due to lack of temperature measurements to ± 33 $\mu\varepsilon$.

The error in modelling elastic strain has different sources. The first source is related to uncertainty of the real load at the time of measurement. Although the measurements were performed after each new storey was built, the real load can be different from the design values

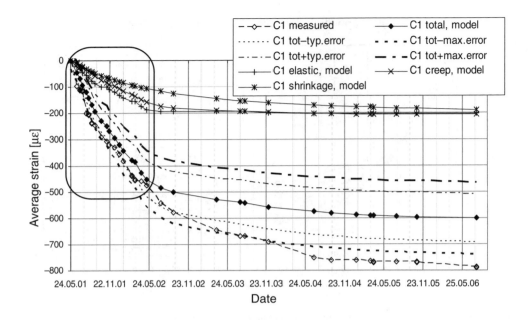

Figure 3.21   Comparison between the numerical model and results of monitoring for column C1 (data provided by HDB and SMARTEC).

due to storage of material on previously built storeys, due to work on the next storey and finally due to redistribution of load over columns that may not necessarily be the same as calculated. The second source of error is related to eccentric positioning of the sensor with respect to the column's centre of gravity. Owing to the rigid connection between the horizontal structural elements (beams and slabs) with the column, some bending may be present in the column. Since the sensor is supposed to measure only the normal (axial) force, the bending will cause 'parasite' effects on the eccentric sensor. Since it is impossible to evaluate the error in elastic strain by other means, the maximal relative error in determination of elastic strain is estimated to 10 % based on experience and structural analysis.

Creep and shrinkage were determined using the Model Code (CEB-FIP, 1990) with no experimental verification. Thus, the errors could not be estimated precisely. Both creep and shrinkage are products of constants and functions; therefore, it is reasonable to consider the relative error rather than the absolute error. Based on experience, the relative error for shrinkage is set to 20 %. The error in creep depends on the error of the elastic strain (10 %), but it also depends on the error in determination of the equivalent creep coefficient, which is considered to be 20 % (the same as for the shrinkage). Consequently, the relative creep error is considered to be 22.4 % (square root of $0.1^2 + 0.2^2$).

The monitored strain is compared with the model in Figure 3.21. The typical and maximal error limits are also presented in the figure.

During the construction phase (the first 12 months) the monitored strain is in agreement with the model. However, during the months that follow, a drift from the model is noticed, and this drift exceeded the typical error 4 months later and maximal error 1.5 years later. Since the monitored strain exceeded the maximal error, the condition in Equation 3.58 is not fulfilled and one of the points (A)–(E) presented in Section 3.3.2 is valid. Based on the results of monitoring obtained on the other columns that are in mutual correlation, one can consider the sensor measurements as reliable. In addition, in the construction phase the monitoring results were in agreement with the model; thus, the model can be considered as appropriate. Certain doubts concerning the determination of the error are present, but the fact that the drift shows a clear tendency indicates that there is an anomaly in structural behaviour. This anomaly is expressed through an increase of the total strain that can be explained by:

1. Overloading of the column (e.g. by an unknown live load)
2. Creep and shrinkage evolution
3. Stiffness of the second-storey three-dimensional structural frame and interaction with the other columns that have not been equipped with sensors
4. Unequal settlements of the foundations in columns and neighbouring cores
5. Inclination (rotation) of the second storey.

The analysis presented in Section 4.2.5 indicates that unequal settlement of neighbouring columns is the most probable reason for the drift. The strain in the column is far below serviceability and ultimate strain limits, and the drift has a tendency to stabilize. Therefore, the anomaly detected does not present a risk for the residents. However, the strain evolution is to be monitored in the future in order to provide the data necessary to trigger an early rehabilitation process. Further analysis of the application above is presented in Sections 4.2.5 and 5.3.

4

Sensor Topologies: Monitoring Global Parameters

4.1 Finite Element Structural Health Monitoring Concept: Introduction

The notion of the deformation sensor's gauge length and its impact on measurement was presented in Chapter 3. An original structural monitoring method is developed in this chapter. The idea is to divide the structure in to parts (elements), called cells (Vurpillot, 1999), to equip each cell with a sensor combination, called a topology (Inaudi and Glišić, 2002b), which in the best manner corresponds to the expected strain field in the cell, and then to link the results obtained from each cell, using appropriate algorithms, in order to retrieve the global structural behaviour. This concept is, by general approach, similar to the concept of the finite element method (FEM), and this is why it is tentatively termed the finite element SHM (FE-SHM) concept (Glišić and Inaudi, 2006). The FE-SHM concept is presented schematically in Figure 4.1.

To perform monitoring at a structural level it is necessary to cover the structure, or a part of it, with sensors. The FE-SHM concept supports rational and efficient instrumentation of the structure. The structure is first divided into cells. Each cell contains a sensor combination appropriate to the parameters representative for this cell (e.g. strain, shear strain, curvature, relative displacement, etc.). Knowing the behaviour of each cell, it is possible to retrieve the behaviour of the entire structure. The combination of sensors installed in a single cell is called the sensor topology. The totality of sensors is called the sensor network.

Deformation sensors can be combined in different topologies and networks, depending on the geometry and type of structure to be monitored, allowing monitoring and determination of important structural parameters such as average strains and curvatures in beams, slabs and shells, average shear strain, deformed shape and displacement, crack occurrence and quantification, as well as damage detection. A sensor network can contain cells with different topologies (see Figure 4.1).

First, four typical topologies of deformation sensors are presented: simple, parallel, crossed and triangular. Then, the algorithms for determining the global structural behaviour are developed and illustrated with on-site examples.

Fibre Optic Methods for Structural Health Monitoring   B. Glišić and D. Inaudi
© 2007 John Wiley & Sons, Ltd

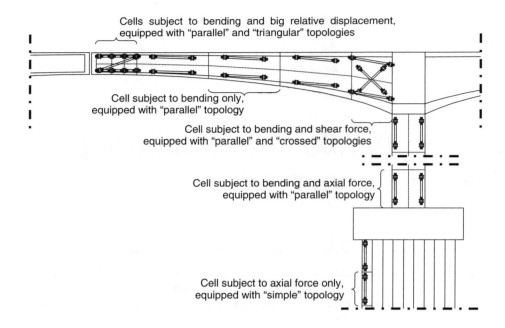

Figure 4.1   Schematic representation of FE-SHM concept (courtesy of SMARTEC).

4.2 Simple Topology and Applications

4.2.1 Basic Notions on Simple Topology

A simple topology (Inaudi and Glišić, 2002b) consists of a single sensor installed by preference
in a direction of principal strain. It is mainly used for monitoring linear structural elements
(beams) subjected to axial compression or traction combined with longitudinal shear stresses
and a dead load (see Figure 4.2), such as piles or columns. In these cases no bending occurs
and the stress and the strain are constant over the cross-section of the beam. Thus, the sensor
can be installed regardless of position in the cross-section (not necessarily in the centre of
gravity) and provide information directly related to the structural behaviour of the cell being
monitored.

An example of a cell (cell i) subject to normal (axial) stresses σ_i and σ_{i-1}, longitudinal
shear stresses (friction) τ_i and dead load g_i, equipped with a simple topology is presented in
Figure 4.2.

The simple topology is used in an enchained or scattered configuration. These configurations
are developed in the following sections. An on-site example of the simple topology with $\tau_i =
0$, applied on a building's columns, was presented in Section 3.3.4 and is further developed in
Section 4.2.4.

The simple topology can also be used in cases when the strain field in a monitored element
is complex, and the principal strain is not in the direction of the sensor. The sensor will provide
information (measure) concerning the average strain in the direction of its gauge length, but

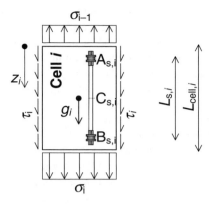

Figure 4.2 Example of a cell with a simple topology (courtesy of SMARTEC).

no direct conclusions concerning the structural behaviour of the monitoring element can be carried out.

4.2.2 Enchained Simple Topology

Linear structural members subject to dominant uniaxial strain, such as piles, are monitored using an enchained simple topology. The structural member is split in to cells, each containing a simple topology. A schematic example of an enchained topology is presented in Figure 4.3.

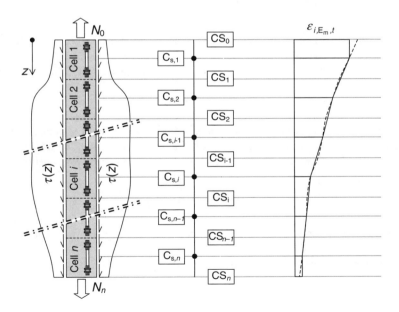

Figure 4.3 Schematic representation of an enchained simple topology.

If several cells containing a simple topology are enchained and fully cover the element being monitored, then the distribution of strain along the element and the relative displacement in the direction of the element can be retrieved. The relative displacement is obtained as the integral sum of strain. In addition, if the Young's modulus and thermal expansion coefficient of the construction material are known, and the rheologic strain (shrinkage and creep) can be estimated, then the distribution of normal forces and longitudinal shear strain (skin friction) can be determined. A simplified method is presented in this section.

Let $\varepsilon_{i,E_m,t}$ be the measurement of elastic strain of the sensor installed in cell i, attributed to the middle point $C_{s,i}$ of the cell (determined from measurement using Equation 3.61, 3.66 or 3.73, depending on the construction material). Then the distribution of the average strain between the cells' middle points can be considered as linear. This simplification introduces the error of the values obtained at the border cross-sections CS_i of the cells, but keeps accurate the values in the cells' middle points and simplifies the analysis (see Figure 4.3).

The relative vertical displacement between the top and bottom cross-sections CS_{i-1} and CS_i of cell i is calculated as follows:

$$\Delta v_{CS_{i-1}-CS_i} = \Delta L_{cell,i} = \varepsilon_{i,m,t} L_{cell,i} = m_{s,i,t} \frac{L_{cell,i}}{L_{s,i}} \tag{4.1}$$

The relative vertical displacement of cross-section CS_i with respect to the bottom cross-section CS_n of the structural member is calculated as presented in Equation 4.2. The bottom cross-section is chosen as a reference since this is expected to be without movement as long as the structural member (pile) is functioning properly.

$$\Delta v_{CS_i-CS_n} = \sum_{j=i+1}^{n} \Delta L_{cell,j} = \sum_{j=i+1}^{n} (\varepsilon_{j,m,t} L_{cell,j}) = \sum_{j=i+1}^{n} \left(m_{s,j,t} \frac{L_{cell,j}}{L_{s,j}} \right) \tag{4.2}$$

The relative displacement of the top of the structural member with respect to the bottom is obtained by inserting $i = 0$ in Equation 4.2.

The normal force in the cells' middle points is determined using the following expression:

$$N_{C_{s,i},t} = E_{i,t} A_{i,t} \varepsilon_{i,E_m,t} \tag{4.3}$$

In the case of reinforced concrete, the Young's modulus $E_{i,t}$ and the area of cross-section $A_{i,t}$ of the cell i can vary in time depending on the age of the concrete and the presence of cracks. For noncracked concrete, the area of the cross-section is calculated using

$$A_{i,t} = A_{i,conc,t} + A_{i,rebar,t} \frac{E_{i,rebar}}{E_{i,conc,t}} \tag{4.4}$$

where indexes 'conc' and 'rebar' indicate concrete and rebar properties respectively ($E_{i,t} = E_{i,conc,t}$ in Expression 4.3).

The potential crack in cell i is detected by the following criterion:

$$\text{cracks} \Leftrightarrow \varepsilon_u^+ \leq \varepsilon_{i,E_m,t} \tag{4.5}$$

where ε_u^+ is the ultimate tensional strain of concrete.

In the case of cracks in uniaxially loaded members, the cracks are fully transverse and the area of the cross-section is reduced to the equivalent area of the cross-section of the rebars, as presented in Equation 4.6:

$$A_{i,t} = A_{i,\text{rebar},t}\,\frac{E_{i,\text{rebar}}}{E_{i,\text{conc},t}} \tag{4.6}$$

The longitudinal shear stress (e.g. skin friction in the case of piles) is assumed to be constant between the cells' middle points and is calculated for each cell from the equilibrium of forces:

$$\tau_{C_{s,i-1}-C_{s,i},t} = 2 \times \frac{N_{C_{s,i},t} - N_{C_{s,i-1},t}}{S_{\text{cell},i} + S_{\text{cell},i-1}} \tag{4.7}$$

where $S_{\text{cell},i}$ is the circumferential surface of cell i.

An on-site example of an enchained simple topology applied to piles is presented in Section 4.2.3.

4.2.3 Example of an Enchained Simple Topology Application

A semiconductor production facility in the Tainan Scientific Park, Taiwan, was to be built on a soil consisting mainly of clay and sand with poor mechanical properties (see Figure 4.4). The

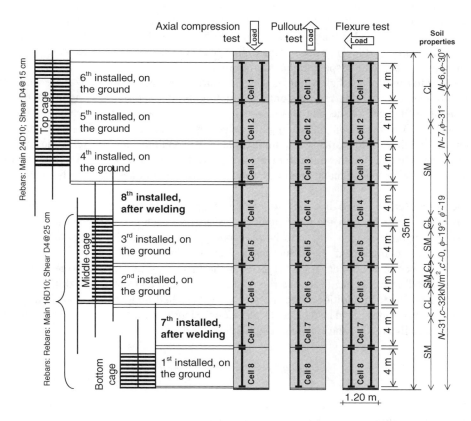

Figure 4.4 Pile dimensions, soil profile, sensor networks and installation sequences on rebar cage (courtesy of SMARTEC).

natural water content in the soil was approximately 20–25 %. Adequate functioning of such a facility is possible only if a high stability of its foundations is guaranteed. It was estimated that approximately 3000 piles would be necessary at that site. To assess the foundation performance, it was decided to perform an axial compression, pullout and flexure test in a full-scale on-site condition. Interferometric fibre-optic long-gauge sensors were used to monitor the behaviour of the six piles under test (Glišić et al., 2002b).

Two sets of reverse, cast-in place piles were tested, one set located on the east side and the other on the west side of the future facility. Each set consisted of three piles, and each pile in a set was tested to a single load case, namely compression (according to ASTM D1143-B1), uplift (according to ASTM D3689-B3) or horizontal force (according to ASTM D3966-90). All piles had the same dimensions (a diameter of 1.20 m and length of 35 m) and were designed and constructed in order to have the same mechanical properties. The compressive strength of 3-week-old concrete samples was 24.5 MPa and the calculated compression and uplift capacity were 365 t and 220 t respectively. The dimensions of the piles, rebar layout and simplified soil mechanical properties are presented in Figure 4.4.

The piles subject to axial compression and pullout test were equipped with an enchained simple topology except in the first cell, where parallel topology (see Section 4.3) was used in order to evaluate the eccentricity of the applied force.

The load was applied stepwise using hydraulic jacks and according to a predetermined programme. The magnitude of the applied load was monitored on the hydraulic scale of the loading setup. Each level of load was maintained during a period whose length was determined depending on the load level. After the maximum load was reached, the piles were unloaded, again stepwise, and with respect to the programme.

In order to monitor the behaviour of the piles during testing, each pile was instrumented with long-gauge fibre-optic sensors combined in appropriate topologies. In addition, the displacement of the head of the pile was recorded using linear variable displacement transducers (LVDTs). The measurement readings were performed immediately after each load step and at several times afterwards, while the load level was maintained. The schedule for loading and the schedule for the measurements are presented in Table 4.1. The numbers in the columns labelled 'measurements' indicate the elapsed time in minutes after reaching the target load.

The full presentation and discussion of each measured parameter largely exceeds the scope of this book; therefore, only the most significant results for each particular test are presented. In this section, only the tests related to simple topology (i.e. compression and pullout tests) are presented. As a complement, the flexure test is presented in Section 4.3.4.

The reference measurement for each test was the first measurement performed. Since the tests were performed within 1 day, the rheologic strain (creep and shrinkage) during the tests could be neglected. In addition, with the pile being installed underground, the influence of temperature variations could also be neglected. Consequently, the total strain measured in each cell is practically equal to the elastic strain; that is:

$$\varepsilon_{i,\mathrm{E_m},t} = \varepsilon_{i,\mathrm{m},t} = \frac{m_{\mathrm{s},i,t}}{L_{\mathrm{s},i}} \tag{4.8}$$

This parameter served as a basis to calculate all other parameters. The distributions of the elastic strain over the length of the pile, in the case of the axial compression test, on the east-side

Table 4.1 Loading and measurement schedule

Step	Pullout test		Axial compression test		Flexure test	
	Load (t)	Measurement at (min)	Load (t)	Measurement at (min)	Load (t)	Measurement at (min)
0	0.0		0		0	
1	28.6	0, 2, 5, 10	60	0, 2, 5, 10	10	0, 3, 6, 10
2	57.1	0, 2, 5, 10	120	0, 2, 5, 10	20	0, 3, 6, 10
3	85.7	0, 2, 5, 10	180	0, 2, 5, 10	30	0, 3, 6, 10, 15, 20
4	114.3	0, 2, 5, 10	240	0, 2, 5, 10	40	0, 3, 6, 10, 15, 20
5	142.9	0, 2, 5, 10	300	0, 2, 5, 10	50	0, 3, 6, 10, 15, 20
6	171.4	0, 2, 5, 10	360	0, 2, 5, 10	60	0, 3, 6, 10, 15, 20
7	200.0	0, 2, 5, 10	420	0, 2, 5, 10	70	0, 3, 6, 10, 15, 20
8	228.6	0, 2, 5, 10	480	0, 2, 5, 10	80	0, 3, 6, 10, 15, 20
9	257.1	0, 2, 5, 10	540	0, 2, 5, 10	90	0, 3, 6, 10, 15, 20
10	285.7	0, 2, 5, 10	600	0, 2, 5, 10	100	0, 3, 6, 10, 15, 20, 40, 60
11	314.3	0, 2, 5, 10	660	0, 2, 5, 10	75	0, 3, 6, 10
12	342.9	0, 2, 5, 10	720	0, 2, 5, 10	50	0, 3, 6, 10
13	371.4	0, 2, 5, 10	780	0, 2, 5, 10	25	0, 3, 6, 10
14	400.0	0, 2, 5,10, 20, 30, 40, 60	840	0, 2, 5, 10, 20, 30, 40, 60	0	0, 3, 6, 10, 15, 20
15	300.0	0, 2, 5, 10	630	0, 2, 5, 10		
16	200.0	0, 2, 5, 10	420	0, 2, 5, 10		
17	100.0	0, 2, 5, 10	210	0, 2, 5, 10		
18	0.0	0, 2, 5, 10, 20, 30, 40, 60	0	0, 2, 5, 10, 20, 30, 40, 60		

pile and for increasing loads are presented in Figure 4.5 and the same type of diagram obtained in the case of the pullout test is presented in Figure 4.6.

In Figure 4.5, a soil layer with poor mechanical properties was identified (encircled area). In this layer the average strain in the pile is constant, indicating that the friction between the pile and the soil is low, that the stiffness/strength of the soil is low or that the cross-section of the pile is restricted. Ultrasonic testing confirmed that the cross-section of the pile did not change; therefore, this anomaly is most probably due to poor mechanical properties of the soil at that particular depth.

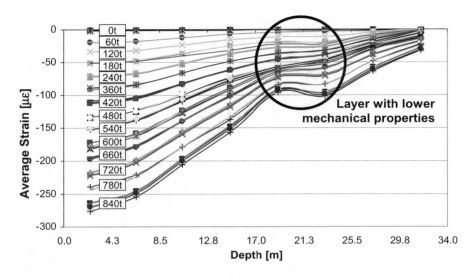

Figure 4.5   Average strain distribution, increase of load, axial compression test, east-side pile (courtesy of SMARTEC, data by RouteAero).

In Figure 4.6, a sudden increase in the strain magnitude was noticed. This was the consequence of crack formation that allowed the detection of damage (cracking) and its propagation along the pile, as a function of the load magnitude. The crack occurred at an elastic strain of approximately 60 $\mu\varepsilon$; thus, in the further analysis, it is considered that $\varepsilon_u^+ = 60\ \mu\varepsilon$. When the maximum load was applied, cracks appeared in the first three cells (see Figure 4.5).

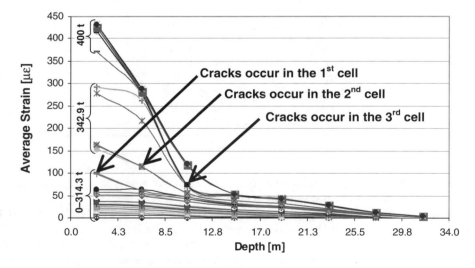

Figure 4.6   Average strain distribution, increase of load, pullout test, east-side pile (courtesy of SMARTEC, data by RouteAero).

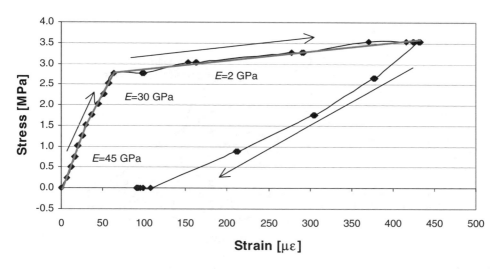

Figure 4.7   Stress–strain diagram for east-side pile under the pullout test (courtesy of SMARTEC, data by RouteAero).

The Young's modulus of the pile is directly determined during the axial compression and pullout tests. During both tests the average strains in the first and second cells were approximately identical for lower loads, which indicated that the soil friction in the first cell could be neglected. Hence, Young's modulus for each pile was calculated as the ratio between applied stress and average strain in the first cell. The stress is calculated as the ratio between the applied force and the area of the cross-section that is considered as constant. This is an alternative method to Equation 4.6, where Young's modulus is kept constant and the area of the cross-section is variable.

The behaviour of piles under the axial compression test was linear, and the value of the Young's modulus ranged from 30 to 50 GPa. The piles subjected to traction had approximately bilinear or trilinear behaviour during loading and nonlinear behaviour during unloading. The stress–strain diagram obtained from measurements on the east-side pile during the pullout test is presented in Figure 4.7. A sudden decrease in Young's modulus was clearly observed after concrete cracking (compare with Figure 4.6).

The average normal (compressive or tensile) force in the pile is determined as the product of the average strain, cross-sectional area and the previously determined Young's modulus. In the case of axial compression tests, with the Young's modulus being constant, the compressive force distribution diagram is proportional to the one presenting the strain distribution presented in Figure 4.5.

For the pullout test, the tensile force distribution diagram is presented in Figure 4.8. It has been determined taking into account the stress–strain dependence presented in Figure 4.7. The soil layer with poor mechanical properties was once again detected. Since its position is slightly different from that determined by the axial compression test (see Figure 4.5), we suppose that this layer is situated in the lower part of the fourth cell and upper part of the fifth cell. This finding also explains the nonconstant value of the force in the pile surrounded by soil with poor mechanical properties.

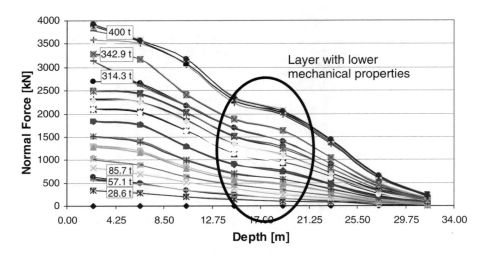

Figure 4.8 Normal force distribution for east-side pile under the pullout test, increase of load (courtesy of SMARTEC, data by RouteAero).

The ultimate load capacity of the pile subjected to compression has been determined as the minimal load causing failure of the lateral friction. It was determined as the load that significantly increases the slope on the tip force versus load diagram. When this slope increases significantly, the pile starts to slip and the tip force is activated. The slippage of the pile was also confirmed by the pile-head displacement measured by the LVDT. The determination of the ultimate load capacity, as well as the bottom force versus load diagram, is presented in Figure 4.9. The bottom force was assumed to be approximately equal to the compressive force in the bottom cell.

In case of the pile exposed to traction, the ultimate uplift capacity is determined as the minimal load that damages the pile. This is clearly observed in Figure 4.6. This load significantly increases the slope on the pile-head vertical displacement versus load diagram, since the cracks open and the stiffness decreases.

The pile-head displacement was measured using the LVDT, but was also calculated from deformations measured by fibre-optic sensors using Equation 4.2, assuming that the pile tip did not move. The pile-head displacement and the determination of the ultimate uplift capacity are presented in Figure 4.10. In this diagram the difference between the LVDT and fibre-optic measurements represents the cracks opening in the head of the pile, in the upper zone not covered with the fibre-optic sensor.

Comparisons between the calculated (predicted) bearing capacities and the ultimate capacities of piles obtained from tests showed that the short-term safety factors were not satisfactory (1.32 instead 2.00 for compression and 1.56 instead of 3.00 for uplift), leading to the conclusion that the piling design was not satisfactory. Both the soil–pile friction surface and the total pile traction strength must be increased.

The calculated distributions of pile–soil friction for axial compression at the ultimate load capacity (480 t) and at the maximum applied load (840 t) are presented in Figure 4.11. Three zones of soil with different mechanical properties are identified and highlighted in this figure.

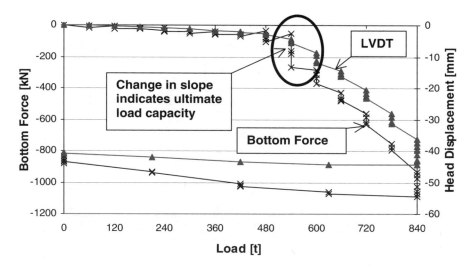

Figure 4.9   Bottom force–load diagram and determination of ultimate load capacity of the pile, east-side pile, axial compression test (courtesy of SMARTEC, data by RouteAero).

While the first and the third zones have good mechanical properties, the mechanical properties of the second zone are inadequate.

The foundation performances determined using the long-gauge fibre-optic sensors are listed in Table 4.2 (data analysis for flexure test is presented in Section 4.3.4). The Young's modulus ranged between 45 and 50 GPa with the exception of the east-side pile tested in compression,

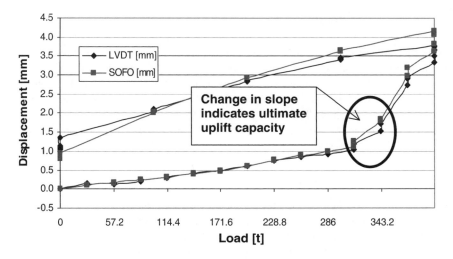

Figure 4.10   Head displacement–load diagram and determination of ultimate uplift capacity of the pile, east-side pile, pullout test (courtesy of SMARTEC, data by RouteAero).

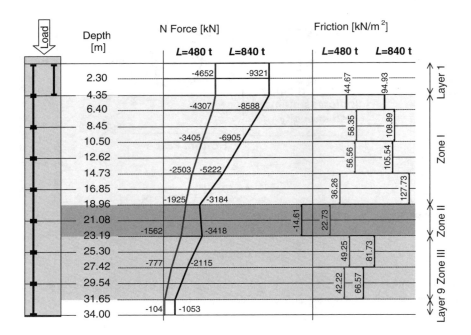

Figure 4.11 Distributions of normal force in the pile, friction stress and different zones of soil (courtesy of SMARTEC, data by RouteAero).

where the Young's modulus was lower (30 GPa). The maximum traction strain of concrete determined during pullout and flexure tests is approximately equal for all piles and corresponds to 60 μɛ. Therefore, the traction strength of the concrete is estimated to be about 2.7–3.0 MPa. Since the mechanical properties were relatively uniform among all piles, it can be concluded that their quality was identical.

The enchained simple topology helps in understanding the behaviour of real piles and in evaluating their performance, in determining failure modes and localizing cracking zones and failure points. In addition, the mechanical properties of the soil were determined and, globally, three layers with different properties were distinguished at both the east and the west sites. The ultimate load capacity for all tests was approximately equal to half of the maximum applied load, which is in agreement with the design values.

In order to determine the Young's modulus and the tip force more accurately, it is recommended to use shorter sensors (1–2 m) in the first and last cells.

4.2.4 Scattered Simple Topology

A scattered simple topology is used for monitoring structures supported by linear elements (e.g. buildings supported by columns). Several columns at the same level (storey) in a building equipped with a simple topology create a scattered simple topology. Examples of the use of simple topology in column monitoring are presented in Sections 1.3 and 3.3.4.

The analysis of a scattered simple topology is not structured as in the case of enchained topology, but rather consists of several algorithms that are to be applied in order to assess

Table 4.2   Foundation performances determined using long-gauge fibre-optic sensors

	Pullout test	Axial compression test	Flexure test
Young's modulus E of pile (GPa)	45–50	30–50	Not calculated
Deformation of pile	Average longitudinal strain distribution Distribution of vertical displacement	Average longitudinal strain distribution Distribution of vertical displacement	Average longitudinal strain distribution Distribution of curvature Distribution of horizontal displacement (deformed shape)
Forces in pile	Distribution of tensile force Bottom force	Distribution of compressive force Bottom force	Qualitative distribution of bending moments
Strain ε when cracks occur ($\mu\varepsilon$)	60	No crack detected	60
Damaging of pile	Detection of crack occurring Localization of zone affected by cracking	No damaging detected	Detection of crack occurring Localization of zone affected by cracking
Properties of soil	Qualitative determination of soil strength Identification of zones with different mechanical properties	Qualitative determination of soil strength Identification of zones with different mechanical properties	Qualitative determination of soil strength
Forces in soil	Distribution of pile-soil friction	Distribution of pile-soil friction	Distribution of horizontal reactions of soil
Failure mode	On pile (cracking)	On soil (slip)	On soil (first) and pile (afterwards)
Ultimate load capacity (t)	314.3–343.2	480–540	50

the global structural behaviour. Supposing the storey slab is provided with high stiffness, the columns supporting the slab are expected to deform for similar absolute values and the total strains in different columns are expected to be in mutual linear correlation. The two criteria presented above are, in general, used to detect imperfections in column-supported structures.

linear correlation between the elastic strains of the columns is lost.

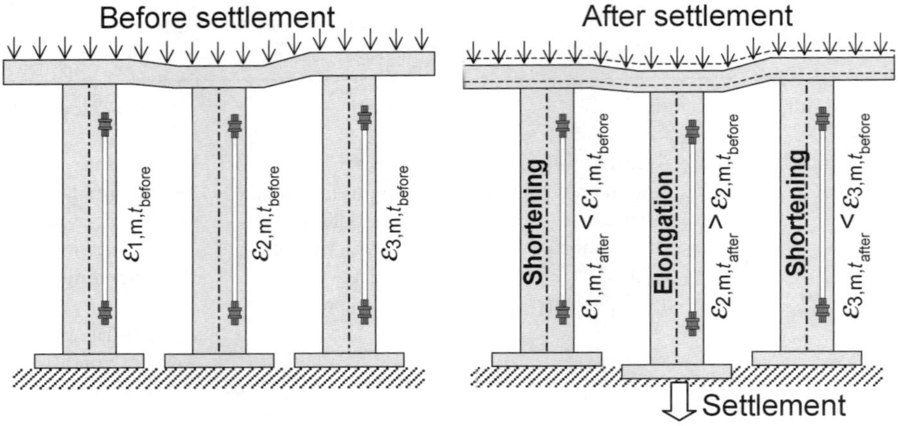

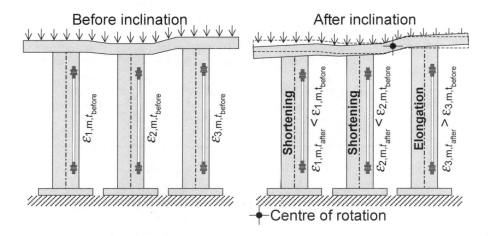

Figure 4.12   Detection of unequal settlement cf foundations (courtesy of SMARTEC).

An example of detection of unequal foundation settlement is presented in Figure 4.12. The column with foundation subject to settlement will elongate while the neighbouring columns will shorten. In this case the linear correlation between the elastic strains of the columns is lost.

Similar to the previous example, an inclination of the building storey as a whole is detected from unequal deformation of the columns, as presented in Figure 4.13. The deformation of columns found on one side of the centre of rotation of the building will systematically decrease while the deformation of the columns found on the other side will increase.

Figure 4.13   Detection of inclination of the storey.

Besides the descriptive analysis presented above, statistical analysis and, in particular, analysis of linear correlation between the measurements of different sensors can ascertain the structural malfunction. An example of analysis of a scattered simple topology is presented in the next section.

4.2.5 Example of a Scattered Simple Topology Application

The example of a scattered simple topology is presented for the Singapore building project whose description is given in Section 3.3.4 (Glišić *et al.*, 2005). The position of columns equipped with sensors is presented in Figure 4.14. This configuration of sensors allows, on a local structural level, the monitoring of columns with different cross-sections and, on a global level, the monitoring of the structural behaviour of four units and an estimation of the global building behaviour.

The results of monitoring are presented in Figure 3.20 and the analysis at local column level is presented in Section 3.3.4. Analysis at the global level is based on comparison between the strains measured in the columns belonging to the same unit and, globally, between all the instrumented columns. The main issues expected are unequal settlement of foundations that may produce redistribution of strains and stresses in columns and in some cases rotations of the second floor. An example of analysis will be performed on Unit A, containing the columns C1, C2 and C3 equipped with sensors (see Figure 4.14). The monitored strain evolutions for these columns are presented in Figure 4.15.

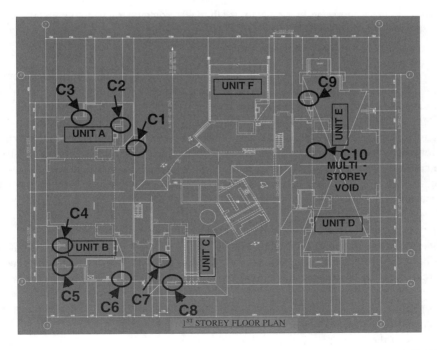

Figure 4.14 Position of the columns in ground floor equipped with sensors (courtesy of SMARTEC, HDB and Sofotec).

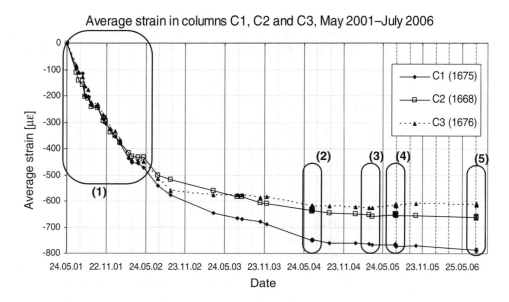

Figure 4.15 Evolution of total average strain in columns C1, C2 and C3, monitored over more than 5 years; highlighted periods: (1) construction of 19 storeys; (2) the first 48 h continuous monitoring session performed in July 2004; (3) before and after tremor monitoring; (4) the second 48 h continuous monitoring session performed in July 2005; (5) the third 48 h continuous monitoring session performed in July 2006 (data provided by HDB and SMARTEC).

The average strain measured in the columns is approximately the same for each column (see Figure 4.15) during the first 12 months, the period that includes the construction. During this period the second-floor slab practically displaced as a rigid body. The analysis at column level presented in Section 3.3.4.2 demonstrated that, during construction, the total strain evolution was in full agreement with numerical models. The fact that the measured strain is approximately equal for each column and in good agreement with the numerical model during the construction period confirms the proper performance of the Unit A during the construction.

After the construction, the columns continue to have similar behaviour for the next 4 months. The measurements are still within the error limits and the behaviour of the Unit A is still sound.

In the whole period that follows, the estimated strain and the forces in columns C1 and C2 are higher than theoretically predicted, whereas in column C3 they are lower (Glišić *et al.*, 2005), but this discrepancy has a tendency to stabilize in time. The observed differences are due to redistribution of stresses and strains, which may be imposed by several factors, such as:

1. Overloading of the column by a live load
2. Creep and shrinkage evolution
3. Stiffness of the second-storey three-dimensional structural frame and interaction with the other columns that have not been equipped with sensors
4. Unequal settlements of the foundations in columns and neighbouring cores

5. Inclination (rotation) of the second storey.

Overloading of the columns as a group is not detected, since the sum of forces in the columns concerned obtained from monitoring is approximately equal to the corresponding sum obtained from the theoretical prediction.

The global analysis in the long term consists of observation of correlation changes between the columns. If no degradation in performance occurs, then the correlation between the columns is expected to be linear, since the horizontal elements, beams and slabs, impose a linear redistribution of total strain. If a malfunction occurs in one column, then the correlation between this column and other columns will no longer be linear. The correlation between column C2 and the other two columns is presented in Figure 4.16. Change in linearity four months after construction is observed in Figure 4.16, in particular for column C3, which deforms less than expected.

Since shrinkage is similar for all columns and creep is proportional to elastic strain, the simple redistribution of stresses and strains will reflect small changes in linearity. Since all the units in the building are interconnected with structural elements, it is not likely to expect an independent rotation of the second-storey slab, unless it is damaged. Visual inspection did not confirm any damage; therefore, the most probable reason for discrepancy in column behaviour is unequal settlement of the foundations.

Although an imperfection in structural behaviour was detected, there is no risk for the building's residents so far, since the total strain in the columns is far below the serviceability and capacity limits. Hence, the scattered simple topology allowed detection of unusual structural behaviour at an early stage and pointed to possible future issues (differential settlement of foundations). The analysis of the application presented above is developed further in Section 5.3.

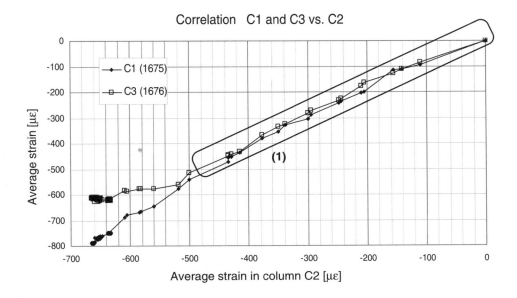

Figure 4.16 Correlation between column C2 and other columns; highlighted period: (1) construction of 19 storeys (data provided by HDB and SMARTEC).

4.3 Parallel Topology

4.3.1 Basic Notions on Parallel Topology: Uniaxial Bending

A parallel topology (Inaudi and Glišić, 2002b) is used for the monitoring of cells subject to bending. It consists of two sensors with equal gauge lengths, parallel to the elastic line of the beam and installed at different levels of the cross-section. An example of a parallel topology is presented in Figure 4.17.

In homogeneous materials, the cross-sectional properties do not change in time and, consequently, the position of the elastic line does not change in time; but in the case of inhomogeneous materials and, in particular, in the case of reinforced concrete, the cross-sectional properties, such as centre of gravity, area and moment of inertia, depend on crack depth and can change in time. Consequently, the elastic line of the beam can change its position. In addition, the viscoelasto-plastic material (concrete) is the most complex, since it involves creep and shrinkage. The case of an inhomogeneous viscoelasto-plastic material is the most general and, therefore, is analysed in more detail in this section. The expressions and considerations for other materials can be obtained by appropriate reduction.

The initial positions of the sensors in a parallel topology (before cracking), the geometrical properties of the cross-section and the coordinate system are presented in Figure 4.17a; the parameters are indexed with 'r' (reference). The distribution of elastic strain and the change in coordinates and displacement of the elastic line e–e due to loads at time t that caused cracking are presented in Figure 4.17b; noncracked concrete is shown in grey and cracked concrete in white in both instances. The point belonging to the neutral axis (elastic strain equal to zero) is denoted by $\eta_{i,t}$, and the corresponding coordinates contain the same letter in the index. The shear force is not presented in Figure 4.17, since it does not affect the measurements of sensors.

The measured average strains in cell i at the positions of sensors 1 and 2 at time t are expressed as follows:

$$\varepsilon_{i1,\mathrm{m},t} = \varepsilon_{i1,\mathrm{E_m},t} + \varepsilon_{i1,\varphi_\mathrm{m},t} + \varepsilon_{i1,\mathrm{T_m},t} + \varepsilon_{i1,\mathrm{sh_m},t} \tag{4.9a}$$

$$\varepsilon_{i2,\mathrm{m},t} = \varepsilon_{i2,\mathrm{E_m},t} + \varepsilon_{i2,\varphi_\mathrm{m},t} + \varepsilon_{i2,\mathrm{T_m},t} + \varepsilon_{i2,\mathrm{sh_m},t} \tag{4.9b}$$

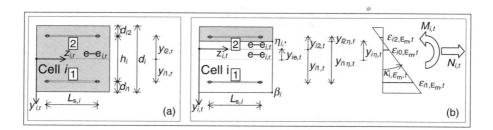

Figure 4.17 Parallel topology in a segment of (a) a noncracked and (b) a cracked concrete beam (courtesy of SMARTEC).

In a parallel topology, the sensors installed at different levels in a cross-section will measure different values of average strain, thus allowing monitoring of average curvature in the cell. The average curvature is calculated assuming that the Bernoulli hypothesis (Brčić, 1989) is satisfied (plane cross-sections of the beam remain plane under loading) using the following expression:

$$\kappa_{i,\mathrm{m},t} = \frac{1}{r_{i,\mathrm{m},t}} = \frac{1}{h_i}(\varepsilon_{i1,\mathrm{m},t} - \varepsilon_{i2,\mathrm{m},t}) = \kappa_{i,\mathrm{E_m}+\varphi_\mathrm{m},t} + \kappa_{i,\mathrm{T_m},t} \qquad (4.10a)$$

where the measured curvature caused by elastic strain E and creep φ is expressed as

$$\kappa_{i,\mathrm{E_m}+\varphi_\mathrm{m},t} = \frac{1}{r_{i,\mathrm{E_m}+\varphi_\mathrm{m},t}} = \frac{1}{h_i}(\varepsilon_{i1,\mathrm{E_m}+\varphi_\mathrm{m},t} - \varepsilon_{i2,\mathrm{E_m}+\varphi_\mathrm{m},t}) \qquad (4.10b)$$

and the measured thermal curvature, supposing that the temperature variation distribution between the sensors is linear, is expressed as

$$\kappa_{i,\mathrm{T_m},t} = \frac{1}{r_{i,\mathrm{T_m},t}} = \frac{1}{h_i}(\varepsilon_{i1,\mathrm{T_m},t} - \varepsilon_{i2,\mathrm{T_m},t}) \qquad (4.10c)$$

Having provided that the temperature is measured and that thermal expansion coefficient is known, the average thermal strain is determined using Equation 3.60 and the average thermal curvature is determined using Equation 4.10c.

The shrinkage, by its nature, has the same value at all points of the cell, including the locations of sensors 1 and 2. Consequently, it contributes only to the central strain and not to curvature ($\varepsilon_{i1,\mathrm{sh}},t = \varepsilon_{i2,\mathrm{sh}},t = \varepsilon_{i0,\mathrm{sh}},t \Rightarrow \kappa_{i,\mathrm{sh}},t = 0$).

Combining Equations 4.10a–4.10c and 3.60 the following expression for measured curvature caused by elastic strain and creep is obtained:

$$\kappa_{i,\mathrm{E_m}+\varphi_\mathrm{m},t} = \frac{\varepsilon_{i1,\mathrm{m},t} - \varepsilon_{i2,\mathrm{m},t}}{h_i} - \alpha_\mathrm{T}\frac{\Delta T_{i1,\mathrm{m},t} - \Delta T_{i2,\mathrm{m},t}}{h_i} \qquad (4.11)$$

Finally, by combining Equations 4.11 and 3.44 the expression for the elastic curvature is obtained:

$$\kappa_{i,\mathrm{E_m},t} = \frac{\kappa_{i,\mathrm{E_m}+\varphi_\mathrm{m},t}}{1 + K_{i,\varphi_\mathrm{m},\mathrm{equ}}f_{i,\varphi_\mathrm{m},\mathrm{equ},t}} = \frac{(\varepsilon_{i1,\mathrm{m},t} - \varepsilon_{i2,\mathrm{m},t}) - \alpha_\mathrm{T}(\Delta T_{i1,\mathrm{m},t} - \Delta T_{i2,\mathrm{m},t})}{h_i(1 + K_{i,\varphi_\mathrm{m},\mathrm{equ}}f_{i,\varphi_\mathrm{m},\mathrm{equ},t})} \qquad (4.12)$$

The equivalent creep coefficient and equivalent creep function are to be determined as presented in Sections 3.1.4 and 3.3.3. The relation between the bending moment and elastic curvature is given for linear elastic relation as follows:

$$M_{i,t} = E_{i,t}I_{i,t}\kappa_{i,\mathrm{E_m},t} = E_{i,t}I_{i,t}\frac{(\varepsilon_{i1,\mathrm{m},t} - \varepsilon_{i2,\mathrm{m},t}) - \alpha_\mathrm{T}(\Delta T_{i1,\mathrm{m},t} - \Delta T_{i2,\mathrm{m},t})}{h_i(1 + K_{i,\varphi_\mathrm{m},\mathrm{equ}}f_{i,\varphi_\mathrm{m},\mathrm{equ},t})} \qquad (4.13)$$

It is important to remember that the Young's modulus E_i of concrete changes in time, as well as the moment of inertial I_i. The change of Young's modulus depends on the concrete's maturity, and for mature concrete it can be considered as approximately constant. Changes in the moment of inertia depend on reinforcement and the crack depth and can be evaluated using the expressions developed in the following text of this section.

The other important parameter to be determined from measurement is the central elastic strain, which is proportional to normal (axial) force. For this purpose it is first necessary to determine the position of the elastic line after cracking. This procedure is very complex and first requires characterization of the cracks.

Assuming that the Bernoulli's hypothesis is valid, the distances from the point $\eta_{i,t}$ belonging to neutral axis (see Figure 4.17) to the sensors are determined by transforming Equation 3.10 as follows:

$$y_{i1\eta,t} = \frac{\varepsilon_{i1,E_m,t}}{K_{i,E_m,t}} = \frac{\varepsilon_{i1,E_m+\varphi_m,t}}{K_{i,E_m+\varphi_m,t}} = \frac{\varepsilon_{i1,E_m+\varphi_m,t}}{\varepsilon_{i1,E_m+\varphi_m,t} - \varepsilon_{i2,E_m+\varphi_m,t}} h_i \qquad (4.14a)$$

$$y_{i2\eta,t} = \frac{\varepsilon_{i2,E_m,t}}{K_{i,E_m,t}} = \frac{\varepsilon_{i2,E_m+\varphi_m,t}}{K_{i,E_m+\varphi_m,t}} = \frac{\varepsilon_{i2,E_m+\varphi_m,t}}{\varepsilon_{i1,E_m+\varphi_m,t} - \varepsilon_{i2,E_m+\varphi_m,t}} h_i \qquad (4.14b)$$

The position of the neutral axis point is determined using

$$y_{i\eta,t} = y_{i1,r} - y_{i1\eta,t} = y_{i2,r} - y_{i2\eta,t} \qquad (4.15)$$

It is important to highlight that determination of the position of the neutral axis point $\eta_{i,t}$ (Equations 4.14 and 4.15) does not depend on the creep and shrinkage parameters. After determining these values it is possible to calculate a number of parameters related to cracks (Glišić and Inaudi, 2002b).

The cracks start to open when the elastic strain at the bottom line of the cross-section β_i achieves the ultimate tensional strain of concrete ε_u^+, as presented in Equation 4.16a. The strain at the bottom line of the cross-section β_i is calculated using a modified Equation 3.10 and geometrical proportion.

$$\text{cracks} \Leftrightarrow \varepsilon_u^+ \leq \varepsilon_{\beta_i,E_m,t} \Leftrightarrow \varepsilon_u^+ \leq \frac{y_{i1\eta,t} + d_{i1}}{y_{i1\eta,t}} \varepsilon_{i1,E_m,t} \qquad (4.16a)$$

In Equation 4.16a, besides the thermal strain, the creep and the shrinkage components must be determined in order to determine the elastic strain $\varepsilon_{i1,E_m,t}$.

The average crack depth at time t is obtained by taking into account the tensioned part of the concrete next to the neutral axis:

$$d_{i,\text{crk},t} = y_{i1\eta,t} + d_{i1,t} - \frac{\varepsilon_u^+}{K_{i,E_m,t}} = y_{i1\eta,t} + d_{i1,t} - \frac{\varepsilon_u^+(1 + K_{i,\varphi,\text{equ}} f_{i,\varphi,\text{equ},t})}{\varepsilon_{i1,E_m+\varphi_m,t} - \varepsilon_{i2,E_m+\varphi_m,t}} h_i \qquad (4.16b)$$

The interpretation of Equation 4.16b is limited in the case the cracks open and then partially close; in that case the depth of the open part of the crack is determined.

The crack width sum along the segment of the bottom line β_i within the length of sensor $L_{s,i}$ is calculated as follows:

$$\sum_{c_j} w_{i,c_j,t} = [\varepsilon_{i1,E_m+\varphi_m,t} - k_w \varepsilon_u^+(1 + K_{i,\varphi,\text{equ}} f_{i,\varphi,\text{equ},t})] \frac{y_{i1\eta,t} + d_{i1}}{y_{i1\eta,t}} L_{s,i} \qquad (4.16c)$$

where k_w is a correction coefficient ($0 \leq k_w \leq 1$) that takes into account the strain in the concrete between two cracks.

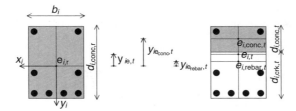

Figure 4.18   Dependence of change of cross-sectional properties on crack depth.

If the number of cracks $n_{i,c,t}$ along the length of sensor is known, then the average opening of the crack at time t is calculated as follows:

$$w_{i,c} = \frac{\sum\limits_{c_j} w_{i,c_j,t}}{n_{i,c,t}} \tag{4.16d}$$

The average depth of the cracks being determined in Equation 4.16b makes possible the determination of the cross-sectional properties of the cracked cell, such as the centre of gravity, the area of cross-section and the moment of inertia. The cross-sectional properties of the cracked cell depend on the concrete cross-section shape, Young's modulus of the concrete and rebars, and the cross-section and geometrical position of rebars, as shown in Figure 4.18.

In Figure 4.18 the reference state is indexed by 'r'. The cross-sectional properties in the reference state in the case of reinforced concrete are calculated as follows:

$$A_{i,r} = A_{i,\mathrm{conc,r}} + A_{i,\mathrm{rebar,r}} \frac{E_{i,\mathrm{rebar}}}{E_{i,\mathrm{conc,r}}} \tag{4.17a}$$

$$I_{i,r} = I_{i,\mathrm{conc,r}} + I_{i,\mathrm{rebar,r}} \frac{E_{i,\mathrm{rebar}}}{E_{i,\mathrm{conc,r}}} \tag{4.17b}$$

$$E_{i,r} = E_{i,\mathrm{conc,r}} \tag{4.17c}$$

where the index 'conc' refers to concrete without reinforcing bars and the index 'rebar' refers to the reinforcing bars.

After cracking, at time t, the concrete area of the cross-section is reduced (grey shading in Figure 4.18) and, consequently, the cross-sectional properties of the concrete (without rebars) are changed as well as its own centre of gravity. The centre of gravity and the cross-sectional properties of the rebars are, in general, unchanged; they can only change in the case of heavy corrosion or breaking of some bars.

The position of the centre of gravity at time t is determined as given in Equation 4.18, and the cross-sectional properties are as given in Equations 4.19a–4.19c.

$$y_{ie,t} = \frac{A_{i,\text{conc},t}\, y_{ie_{\text{conc}},t} + A_{i,\text{rebar},t}\, \dfrac{E_{i,\text{rebar}}}{E_{i,\text{conc},t}}\, y_{ie_{\text{rebar}},t}}{A_{i,\text{conc},t} + A_{i,\text{rebar},t}\, \dfrac{E_{i,\text{rebar}}}{E_{i,\text{conc},t}}} \tag{4.18}$$

$$A_{i,t} = A_{i,\text{conc},t} + A_{i,\text{rebar},t}\, \frac{E_{i,\text{rebar}}}{E_{i,\text{conc},t}} \tag{4.19a}$$

$$I_{i,t} = I_{i,\text{conc},t} + y_{ie,t}^2 A_{i,\text{conc},t} + (I_{i,\text{rebar},t} + y_{ie,t}^2 A_{i,\text{rebar},t})\, \frac{E_{i,\text{rebar}}}{E_{i,\text{conc},t}} \tag{4.19b}$$

$$E_{i,t} = E_{i,\text{conc},t} \tag{4.19c}$$

The values of Young's modulus and the moment of inertia determined in Equations 4.19a–4.19c are used in Equation 4.13 to determine the bending moment in the cell. The central elastic strain and the normal (axial) force (see Figure 4.17) are determined in Equations 4.20 and 4.21 respectively:

$$\varepsilon_{i0,E_m,t} = K_{i,E_m,t}(y_{ie,t} - y_{i\eta,t}) = \frac{(\varepsilon_{i1,m,t} - \varepsilon_{i2,m,t}) - \alpha_T(\Delta T_{i1,m,t} - \Delta T_{i2,m,t})}{h_i(1 + K_{i,\varphi_m,\text{equ}}\, f_{i,\varphi_m,\text{equ},t})}(y_{ie,t} - y_{i\eta,t}) \tag{4.20}$$

$$\begin{aligned} N_{i,t} &= E_{i,t} A_{i,t} \varepsilon_{i0,E_m,t} \\ &= E_{i,t} A_{i,t} \frac{(\varepsilon_{i1,m,t} - \varepsilon_{i2,m,t}) - \alpha_T(\Delta T_{i1,m,t} - \Delta T_{i2,m,t})}{h_i(1 + K_{i,\varphi_m,\text{equ}}\, f_{i,\varphi_m,\text{equ},t})}(y_{ie,t} - y_{i\eta,t}) \end{aligned} \tag{4.21}$$

The analysis of the cell equipped with parallel topology is much simpler if the normal force is not present. In this case the centre of gravity of the cross-section $e_{i,t}$ (elastic line) coincides with the neutral axis point $\eta_{i,t}$; that is:

$$N_{i,t} = 0 \Rightarrow \varepsilon_{i0,E_m,t} = 0 \Rightarrow y_{ie,t} = y_{i\eta,t} \tag{4.22}$$

If the normal force is not present (Equation 4.22 is valid), then the cracks can be detected from the movement of the neutral axis point $\eta_{i,t}$ (which coincides with the centre of gravity $e_{i,t}$):

$$\text{cracks} \Leftrightarrow y_{i1\eta,t} \neq y_{i1,r} \wedge y_{i2\eta,t} \neq y_{i2,r} \Leftrightarrow y_{i\eta,t} \neq 0 \tag{4.23}$$

The advantage of Equation 4.23 with respect to Equation 4.16a lies in fact that it does not depend on the creep and shrinkage parameters, and it is not necessary to know the ultimate tensional strain of the concrete. Actually, if the normal force is not present, then the ultimate tensional strain of the concrete can be approximately determined as the strain at the bottom point of the cross-section at the time t_c^- immediately before the cracks occurred:

$$\varepsilon_{u_m}^+ \approx \frac{y_{i1\eta,t} + d_{i1}}{y_{i1\eta,t}} \, \varepsilon_{i1,E_m,t_c^-} \tag{4.24}$$

The accuracy of Equation 4.24 depends on the elastic strain rate between two measurements. The ultimate tensional strain value determined in Equation 4.24 can be used for crack characterization in Equations 4.16b–4.16d.

4.3.2 Basic Notions on Parallel Topology: Biaxial Bending

The case of a cell subject to uniaxial bending was presented in Section 4.3.1 and this is extended to biaxial bending in this section. Let cell i be subject to biaxial bending and a normal force as presented in Figure 4.19. Cracked concrete is shown in white and noncracked concrete is shown in grey. The Bernoulli hypothesis is supposed to be valid.

The analysis of a biaxially bent cell is more complex for concrete than for other materials, since the cracking may, as a consequence, not only change the position of the centre of gravity (translation of principal axes), the area of the cross-section and the moments of inertia with respect to the principal (main) axes, but also the principal (main) axes may rotate. The translation and rotation of the principal axes do not influence the curvature analysis; therefore, the curvature analysis presented below is carried out with respect to a local reference coordinate system having its origin in reference centre of gravity $e_{i,r}$ without taking into account translation and rotation of the principal (main) axes.

All the sensors used in a parallel topology are to be parallel to the elastic line e–$e_{i,r}$. The minimum number of sensors to be installed in the cell subject to biaxial bending is three, but for reasons of redundancy and accuracy of monitoring it is recommended to install four sensors, as shown in Figure 4.19.

Moreover, in order to simplify the analysis of measurements, it is recommended to select the positions of the sensors in the cross-section symmetrically with respect to the principal (main) axes or centre of gravity, as shown in Figures 4.19 and 4.20.

Let sensors $s_{1,i}$, $s_{2,i}$ and $s_{3,i}$ be installed in a biaxially loaded cell i. Taking into account the considerations about temperature gradients and shrinkage developed in Section 4.3.1, the curvatures $\kappa_{ix,E_m+\varphi_m,t}$ and $\kappa_{iy,E_m+\varphi_m,t}$ with respect to axes $x_{i,r}$ and $y_{i,r}$ are determined from

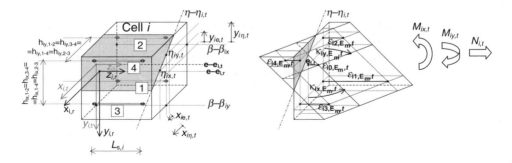

Figure 4.19　Example of a parallel topology, biaxial bending.

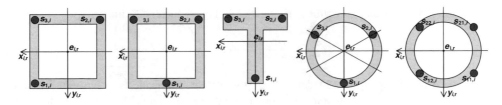

Figure 4.20 Example of recommended cross-sectional positions of sensors in parallel topology, biaxial bending.

the following systems of equations:

$$\kappa_{ix,E_m+\varphi_m,t} h_{ix,1-2} - \kappa_{iy,E_m+\varphi_m,t} h_{iy,1-2} = (\varepsilon_{i1,m,t} - \varepsilon_{i2,m,t}) - \alpha_T(\Delta T_{i1,m,t} - \Delta T_{i2,m,t}) \left.\right\}$$
$$\kappa_{ix,E_m+\varphi_m,t} h_{ix,1-3} - \kappa_{iy,E_m+\varphi_m,t} h_{iy,1-3} = (\varepsilon_{i1,m,t} - \varepsilon_{i3,m,t}) - \alpha_T(\Delta T_{i1,m,t} - \Delta T_{i3,m,t})$$

$$(4.25)$$

where $h_{ix,1-2} = y_{i1,r} - y_{i2,r}$, $h_{iy,1-2} = x_{i1,r} - x_{i2,r}$, $h_{ix,1-2} = y_{i1,r} - y_{i2,r}$, $h_{iy,1-2} = x_{i1,r} - x_{i2,r}$ and $x_{ij,r}$ and $y_{ij,r}$ are the coordinates of sensor s_j, $j \in \{1, 2, 3\}$.

The elastic curvatures are determined as follows:

$$\kappa_{ix,E_m,t} = \frac{\kappa_{ix,E_m+\varphi_m,t}}{1 + K_{i,\varphi_m,equ} f_{i,\varphi_m,equ,t}} \tag{4.26a}$$

$$\kappa_{iy,E_m,t} = \frac{\kappa_{iy,E_m+\varphi_m,t}}{1 + K_{i,\varphi_m,equ} f_{i,\varphi_m,equ,t}} \tag{4.26b}$$

And the thermal curvatures are determined from the following system of equations:

$$\kappa_{ix,T,t} h_{ix,1-2} - \kappa_{iy,T,t} h_{iy,1-2} = \alpha_T(\Delta T_{i1,m,t} - \Delta T_{i2,m,t}) \left.\right\}$$
$$\kappa_{ix,T,t} h_{ix,1-3} - \kappa_{iy,T,t} h_{iy,1-3} = \alpha_T(\Delta T_{i1,m,t} - \Delta T_{i3,m,t})$$

$$(4.27a)$$

The solutions of the equations presented in Equations 4.25 and 4.27a are very simple if the cross-sectional positions of the sensors are selected as recommended in Figure 4.20.

In the case where four (or more) sensors are installed in cell i, the system in Equation 4.25 consists of three (or more) equations (three or more independent sensors' differences) with only two unknowns (curvatures $\kappa_{ix,E_m+\varphi_m,t}$ and $\kappa_{iy,E_m+\varphi_m,t}$) and, in general, may not have the solution. The recommended approach is to use an approximate solution based on the least-squares method.

The neutral axis η-$\eta_{i,t}$ is defined with points $\eta_{ix,t}$ and $\eta_{iy,t}$ belonging to the reference axes; that is, with their coordinates $x_{i\eta,t}$ and $y_{i\eta,t}$ in a referent coordinate system (see Figure 4.19). For any $j \in \{1, 2, 3\}$, the elastic strain in the referent centre of gravity $e_{i,r}$ is determined from the plane of strain as follows:

$$\varepsilon_{ie_r,E_m+\varphi_m,t} = \varepsilon_{ij,E_m+\varphi_m,t} - (\kappa_{ix,E_m+\varphi_m,t} x_{ij,r} - \kappa_{iy,E_m+\varphi_m,t} y_{ij,r}) \tag{4.27b}$$

and coordinates $x_{i\eta,t}$ and $y_{i\eta,t}$ of points $\eta_{ix,t}$ and $\eta_{iy,t}$ are then determined using the following equations:

$$x_{i\eta,t} = \frac{\varepsilon_{ie_r,E_m,t}}{\kappa_{iy,E_m,t}} = \frac{\varepsilon_{ie_r,E_m+\varphi_m,t}}{\kappa_{iy,E_m+\varphi_m,t}} \qquad (4.28a)$$

$$y_{i\eta,t} = -\frac{\varepsilon_{ie_r,E_m,t}}{\kappa_{ix,E_m,t}} = -\frac{\varepsilon_{ie_r,E_m+\varphi_m,t}}{\kappa_{ix,E_m+\varphi_m,t}} \qquad (4.28b)$$

Equations 4.28a and 4.28b do not depend on the creep and shrinkage parameters.

Once the position of the neutral axis is determined, it is possible to determine the cross-sectional parameters of the cell at time t, such as position of the centre of gravity, position of principal (main) axes, area of the cross-section and moments of inertia with respect to the principal axes, based on theory developed in the *Strength of Materials* (Brčić, 1989). Finally, when the cross-sectional parameters are determined it is possible to determine the normal force and bending moments with respect to the principal axes and perform crack characterization, similar to the example presented in Section 4.3.1.

4.3.3 Deformed Shape and Displacement Diagram

If the monitored part of structure contains a representative number of cells equipped with parallel topology, then the average curvature can be monitored in each cell and, consequently, the distribution of curvature over the entire monitored part of the structure can be retrieved. Since the curvature is directly proportional to the bending moment, the distribution of curvature helps to determine the distribution of bending moments. Besides the distribution of bending moments, the distribution of curvature helps in determining the deformed shape of the structure and the diagram of displacements perpendicular to the elastic line.

In order to simplify the presentation, the case of uniaxial bending is analysed and later generalized to biaxial bending. An example of a deformed beam is presented in Figure 4.21.

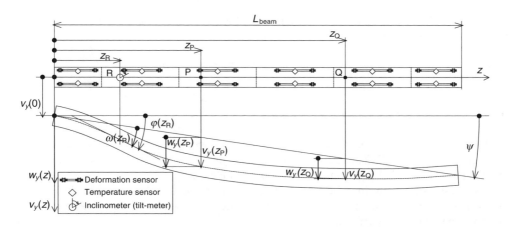

Figure 4.21 Example of deformed beam and notation of diagram of vertical displacements and deformed shape.

The diagram of (absolute) displacements v_y and deformed shape w_y are functions of coordinate z, as shown in the figure.

The deformed shape is defined as a diagram of relative displacements with respect to the axis connecting the extremities of the deformed beam; thus, the condition presented in Equation 4.29 is fulfilled by definition.

$$w_y(0) = w_y(L_{beam}) = 0 \qquad (4.29)$$

The following relations between the diagram of vertical displacement and deformed shape have been established:

$$w_y(z) = v_y(z) - \psi z - v_y(0) \qquad (4.30a)$$

$$\omega(z) = \frac{dw_y(z)}{dz} = \frac{d(v_y(z) - \psi z - v_y(0))}{dz} = \varphi(z) - \psi \qquad (4.30b)$$

$$\kappa(z) = \frac{d^2 w_y(z)}{dz^2} = \frac{d^2(v_y(z) - \psi z - v_y(0))}{dz} = \frac{d^2 v_y(z)}{dz^2} \qquad (4.30c)$$

The angles ψ, ω and φ are considered as small (theory of the first order) and, consequently, they are considered as approximately equal to their tangents in Equations 4.30a–4.30c.

According to Equation 4.30c, the deformed shape is obtained as a double integral of the curvature:

$$w_{y,m,t}(z) = \int \left(\int \kappa_{m,t}(\zeta) \, d\zeta \right) d\zeta + C_1 z + C_2 = W_{y,m,t}(z) + C_1 z + C_2 \qquad (4.31)$$

where

$$W_{y,m,t}(z) = \int \left(\int \kappa_{m,t}(\zeta) \, d\zeta \right) d\zeta \qquad (4.32a)$$

and C_1 and C_2 are constants of integration. The constants of integration are obtained from the boundary conditions given in Equation 4.29 as follows:

$$w_{y,m,t}(0) = 0 \Rightarrow C_2 = -W_{y,m,t}(0) \qquad (4.32b)$$

$$w_{y,m,t}(L_{beam}) = 0 \Rightarrow C_1 = \frac{W_{y,m,t}(0) - W_{y,m,t}(L_{beam})}{L_{beam}} \qquad (4.32c)$$

Let the beam subject to bending be split into cells equipped with parallel topologies, as shown in Figures 4.21 and 4.22. The result of monitoring at time t is a distribution of curvatures $\kappa_{i,m,t}$ along the elastic line of the beam, as shown in Figure 4.22.

As a result of monitoring, the values of curvatures along the elastic line of the beam are known only at a finite number of points; therefore, Equations 4.32a–4.32c are to be calculated using some numerical integration methods.

If the load pattern along the beam is known, then the curvature distribution can be obtained using an appropriate polynomial interpolation (Vurpillot, 1999). For example, if the

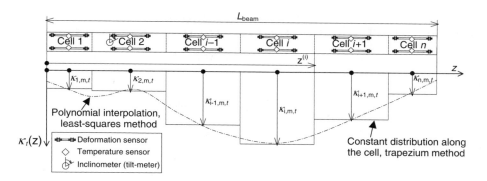

Figure 4.22   Monitored curvature distribution along the beam and different manners of curvature interpolation.

beam is loaded only at extremities, then the curvature distribution along the elastic line is expected to be linear, whereas in the case of a uniformly distributed force the expected curvature distribution is parabolic (see Figure 3.5). The explicit function describing the curvature distribution as a function of coordinate z is obtained by polynomial interpolation and double integration is then performed using the mathematical rules for the integration of polynomials.

For beams subject to different types of load along their elastic line, the beam is to be split in to sub-elements, called macro-elements (Vurpillot, 1999), and each macro-element is to be interpolated with its own polynomial function. The integration constants for each macro-element are then determined from continuity conditions: the deformed shape and its first derivation must be equal for both macro-elements at their common border point. For example, the girder presented in Figure 3.5 is to be split in to five macro-elements delimited with points P_2 to P_5. Macro-element P_1–P_2 is not subject to bending; thus, it is not necessary to monitor it with a parallel topology, except in order to detect eccentricity of the normal force. In this case a single parallel topology is sufficient. Macro-element P_2–P_3 is not loaded between the border points; thus, the expected curvature distribution is linear and at least two cells with parallel topology are necessary. The same discussion is valid for elements P_3–P_4 and P_4–P_5. Finally, the element P_5–P_6 is subject to a uniformly distributed force and the curvature distribution along its elastic line is expected to be parabolic; consequently, a minimum of three parallel topologies are necessary to monitor it. The unknown constants of integration are obtained from the continuity conditions at points P_2–P_5.

In the case of polynomial interpolation, it is recommended to use more topologies than the necessary minimum for each element in order to provide for certain redundancy and increase the accuracy of monitoring. If the number of cells exceeds the minimum necessary number of points for exact polynomial interpolation (e.g. three cells used for a cell with an expected linear distribution of curvatures) then the interpolation polynomial is determined either by using the least-squares method or by using a polynomial with a higher degree.

If the load pattern along the beam is not known, then the curvature can be assumed to be constant over each cell and integration can be performed using the trapezium method:

$$\omega_{m,t}(0) = 0 \tag{4.33a}$$

$$\omega_{m,t}(z^{(i)}) = \sum_{j=1}^{i} \kappa_{j,m,t} L_{\text{cell},j} \tag{4.33b}$$

$$W_{y,m,t}(0) = 0 \tag{4.33c}$$

$$W_{y,m,t}(z^{(i)}) = \frac{1}{2} \sum_{j=1}^{i} \left[(\omega_{m,t}(z^{(j-1)}) + \omega_{m,t}(z^{(j)})) L_{\text{cell},j} \right] \tag{4.33d}$$

$$w_{y,m,t}(z^{(i)}) = W_{y,m,t}(z^{(i)}) - \frac{z^{(i)}}{L_{\text{beam}}} W_{y,m,t}(L_{\text{beam}}) = \frac{1}{2} \sum_{j=1}^{i} \left[(\omega_{m,t}(z^{(j-1)}) + \omega_{m,t}(z^{(j)})) L_{\text{cell},j} \right]$$

$$- \frac{z^{(i)}}{2L_{\text{beam}}} \sum_{j=1}^{n} \left[(\omega_{y,m,t}(z^{(j-1)}) + \omega_{m,t}(z^{(j)})) L_{\text{cell},j} \right] \tag{4.33e}$$

The deformed shape between the points with coordinates $z^{(i-1)}$ and $z^{(i)}$ can be obtained using linear interpolation.

The total measured curvature is used in Equations 4.33a–4.33e and the total deformed shape of the beam is determined from these expressions. To calculate the deformed shape due to temperature variation or loads (elastic with or without creep) only, it is necessary to monitor temperature using temperature sensors. The deformed shape related to each particular influence is obtained by inserting the corresponding monitored curvature, namely the thermal curvature $\kappa_{i,T_m,t}$, the elastic curvature with creep $\kappa_{i,E_m+\varphi_m,t}$ or the elastic curvature without creep $\kappa_{i,E_m,t}$ in Equation 4.33b instead of the total monitored curvature $\kappa_{i,m,t}$. The recommended positions for temperature sensors are presented in Figures 4.21 and 4.22. If the temperature is not expected to vary significantly along the structure, then it is not necessary to equip each cell with temperature sensors. The temperature variation in a cell with no temperature sensor is calculated using linear interpolation between the closest cells equipped with temperature sensors.

The deformed shape is a consequence of straining of the structure and, consequently, it is determined from deformation sensor measurements. In order to determine the diagram of displacements, it is necessary to determine the angle ψ and the displacement of a point with coordinate $z = 0$ (see Equation 4.30a). They can be determined if two characteristics related to absolute displacement are monitored: absolute displacements at two points, P and Q for example (see Figure 4.21). The angle ψ and displacement of the point with coordinate $z = 0$ are calculated as follows, if absolute displacements points P and Q are monitored:

$$\psi_{m,t} = \frac{(v_{y,m,t}(z_Q) - w_{y,m,t}(z_Q)) - (v_{y,m,t}(z_P) - w_{y,m,t}(z_P))}{z_Q - z_P} \tag{4.34a}$$

$$v_{y,m,t}(0) = v_{y,m,t}(z_P) - w_{y,m,t}(z_P) - \psi_{m,t} z_P = v_{y,m,t}(z_Q) - w_{y,m,t}(z_Q) - \psi_{m,t} z_Q \tag{4.34b}$$

If absolute displacement in a point P and absolute rotation in one point R are known, then the angle ψ and the displacement of the point with coordinate $z = 0$ are calculated using Equations 4.35 and 4.34b.

$$\psi_{m,t} = \varphi_{m,t}(z_R) - \omega_{m,t}(z_R) \qquad (4.35)$$

Absolute displacements can be monitored using optical fibre or traditional extensometers, global positioning system (GPS)-based systems or traditional surveying instruments. Absolute rotations can be monitored using optical fibre or traditional inclinometers (tilt-meters). In order to facilitate calculation in Equations 4.34a, 4.34b and 4.35 it is recommended to install the absolute displacement and rotation sensors at the points delimiting cells (i.e. at the points with coordinates $z^{(i)}$).

For biaxially bent beams, the procedures presented in this section are to be repeated for both directions. The application examples of parallel topology for both uniaxial and biaxial bending are presented in the next section.

4.3.4 Examples of Parallel Topology Application

Two application examples of parallel topology are briefly presented in this section. The first example focuses on crack characterization, deformed shape monitoring and displacement monitoring in a pile subject to uniaxial bending (Glišić and Inaudi, 2002b; Glišić et al., 2002b). The second example is related to the deformed shape and displacement monitoring in a biaxially bentbridge (Inaudi et al., 1999a).

A detailed description of the first application example is given in Section 4.2.3, including the load schedule and sensor positions in the pile. In that section, the axial compression and pullout tests were analysed. In this section, the results of the flexure test are presented.

During the flexure test, the order of magnitude of the average strain varied between 0 $\mu\varepsilon$ in cell 8 and 1000 $\mu\varepsilon$ in cell 2. The average strain with respect to load is shown in Figure 4.23.

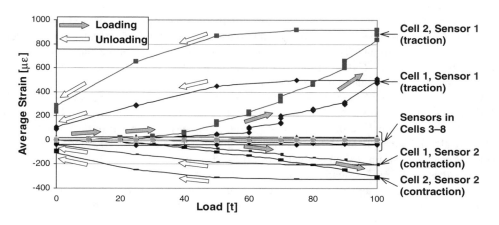

Figure 4.23 Average strain with respect to load, flexure test, west-side pile (courtesy of SMARTEC, data by RouteAero).

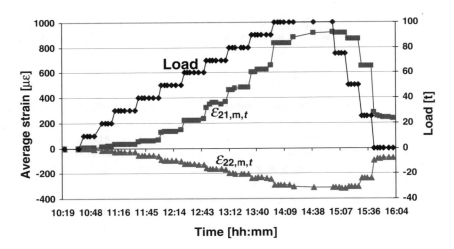

Figure 4.24 Evolution of load applied on the head of the pile and measured strains in cell 2 (courtesy of SMARTEC, data by RouteAero).

For practical reasons, the average strain distribution is shown with respect to the load, not with respect to the depth of the sensor as in the case of axial compression and the pullout test (see Figures 4.5 and 4.6).

A high difference in the strain magnitude for the different cells can be observed in Figure 4.23. Cell 2 was the most deformed, followed by cells 1 and 3; cells 4–8 were practically unaffected, even for the maximum applied load.

For loads below 50 t, parallel sensors installed in each cell measured approximately the same absolute value of deformation; that is, the pile was not cracked for those loads. For higher levels of load, an asymmetry is observed due to cracking and indicating a displacement of the neutral axis. The highest strain level is detected in cell 2, and crack characterization for this cell is performed.

Owing to the short-term duration of the test, the rheologic components of strain (creep and shrinkage) can be neglected. The thermal component can be neglected as well, since the cell is placed 4 m underground. Thus, the monitored total strain consisted of the elastic and plastic strain components only. The evolutions of the load and the average strains measured by sensors 1 and 2 in cell 2 are presented in Figure 4.24.

During the test, the pile was not subject to axial loading and only bending is present in cell 2. Consequently, Equations 4.23 and 4.24 can be used for the crack detection and evaluation of the ultimate tensional strain of the concrete. The crack was detected from the movement of the neutral axis (see Equation 4.23) at approximately $t_c^- \approx 12{:}00$, caused by load of 50 t. The ultimate tensional strain is then evaluated as 60 $\mu\varepsilon$ (Equation 4.24). Crack detection and the ultimate tensional strain evaluation is presented in Figure 4.25.

The evolution of average crack depth is calculated using Equation 4.16b and is presented in Figure 4.26. As expected, the maximum crack opening occurred for maximum load, and the cracks remained open after the load was removed.

The crack width sum is calculated using Equation 4.16c and is presented in Figure 4.27 for different values of the coefficient k_w.

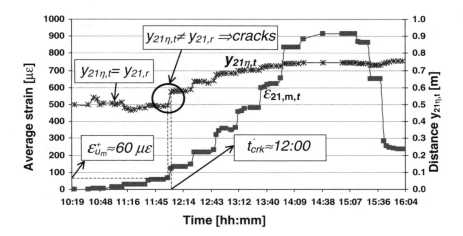

Figure 4.25 Detection of cracks and ultimate tensional strain of concrete in cell 2 (courtesy of SMARTEC, data by RouteAero).

The coefficient k_w depends of the strain level in the concrete and, therefore, is not constant; hence, it cannot be determined accurately. Anyhow, a value of 0.5 can be used as mean value, and values 1.0 and 0.0 set as the lower and upper limits. Since the pile was below the earth surface, it was not possible to determine visually the number of cracks over the length of the sensor; consequently, it was not possible to calculate average crack width according to Equation 4.16d.

The average curvature in the cells was calculated using Equations 4.10a and 4.10b and for the first four cells presented in Figure 4.28. The curvature of the fourth cell can practically be neglected. The same is valid for cells 5–8, and that is why the average curvature for these cells is not presented in the figure.

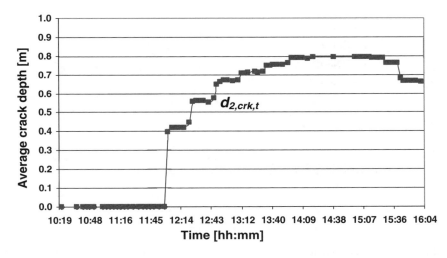

Figure 4.26 Evolution of average crack depth in cell 2 (courtesy of SMARTEC, data by RouteAero).

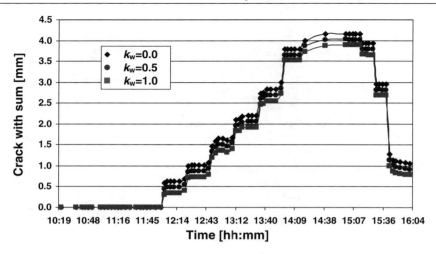

Figure 4.27 Evolution of crack width sum in cell 2 for different values of coefficient k_w (courtesy of SMARTEC, data by RouteAero).

The deformed shape of the pile was calculated using a double integration of the curvature function interpolated with a third-degree polynomial. The third degree was selected assuming that the soil reactions generate linear distribution of loads along the pile. The diagram of horizontal displacements was determined assuming that the horizontal displacement and rotation at the bottom of the pile are equal to zero. This assumption is supported by the fact that the curvature monitored in the bottom cell was practically zero. The horizontal displacement diagram obtained is presented for loads up to 50 t in Figure 4.29 and for loads up to 100 t in Figure 4.30.

The maximum displacement is observed in the first three cells of the pile. The point with maximum curvature in Figures 4.29 and 4.30 corresponds to the failure point of the pile (plastic hinge).

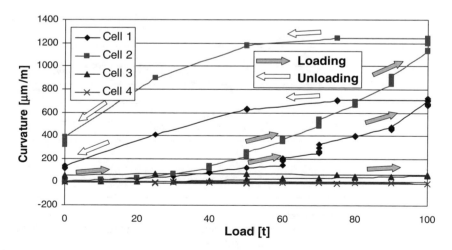

Figure 4.28 Average curvature with respect to load, flexure test, west-side pile (courtesy of SMARTEC, data by RouteAero).

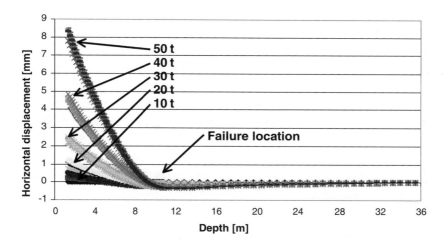

Figure 4.29 Displacement diagram of the pile for loads up to 50 t (courtesy of SMARTEC, data by RouteAero).

As a control for the diagrams determined, the displacement of the head of the pile measured directly with an LVDT is compared with the displacement obtained by double integration of curvatures obtained with parallel topologies of fibre-optic sensors and shown in Figure 4.31. With error smaller than 1 mm, the comparison shows very good agreement between two measurement methods and validates the displacement diagrams obtained.

The ultimate lateral load capacity of the pile was identified as the minimum load that generates cracking in the pile. According to Figures 4.25–4.27 and 4.29–4.30, this load is situated between 40 and 50 t. The pile failed at a depth of approximately 10 m (see Figures 4.29–4.30). The results obtained from the flexure test are summarised in Table 4.2 in Section 4.2.3.

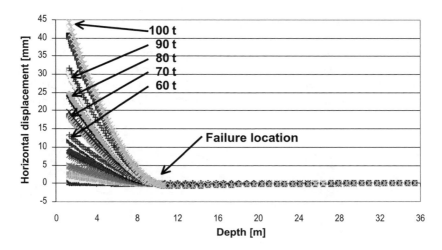

Figure 4.30 Displacement diagram of the pile for loads above 50 t (courtesy of SMARTEC, data by RouteAero).

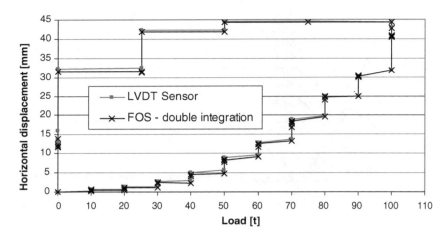

Figure 4.31 Comparison between the displacements of the head of the pile measured directly with an LVDT and determined by double integration of curvatures obtained with parallel topologies of fibre-optic sensors (FOS) (courtesy of SMARTEC, data by RouteAero).

The second application example is related to a bridge subject to biaxial bending. The North and South Versoix bridges are two parallel twin bridges (Vurpillot, 1999). Each one supports two lanes of the Swiss national highway A9 between Geneva and Lausanne. The bridges are classical ones consisting of two parallel prestressed concrete beams supporting a 30 cm concrete deck and two overhangs. In order to support a third traffic lane and a new emergency lane, the exterior beams were widened and the overhangs extended. Because of the added weight and prestressing, as well as the differential shrinkage between new and old concrete, the bridge bends (both horizontally and vertically) and twists during the construction phases. In order to optimize the concrete mix and to increase knowledge on the long-term behaviour and performance, the bridge was instrumented with more than hundred long-gauge fibre-optic sensors. The positions of the sensors in the cross-section are presented in Figure 4.32. A parallel topology is used for monitoring in the horizontal and vertical planes.

The number of the sensors in the cross-section was more than three; thus, the horizontal and vertical curvatures were determined using the least-squares method. The horizontal deformed shape and the diagram of vertical displacements of the first two spans of the bridge were calculated using the double-integration algorithm. Figure 4.33 shows the horizontal deformed

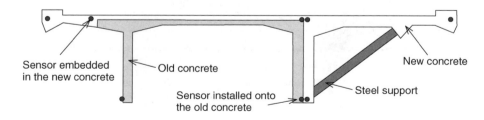

Figure 4.32 Position of sensors in cross-section of Versoix bridge (Vurpillot, 1999) .

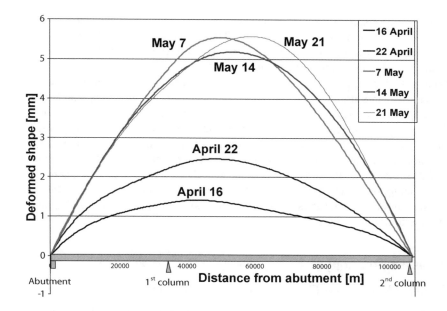

Figure 4.33   Deformed shape of Versoix bridge in the horizontal plane (Vurpillot, 1999) .

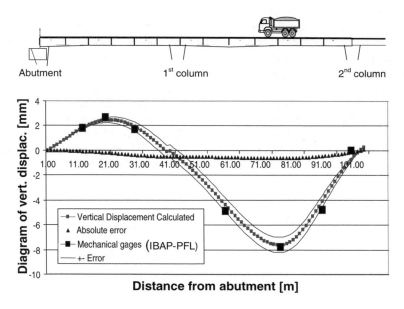

Figure 4.34   Vertical displacement during the load test and comparison with dial gauges (Vurpillot, 1999) .

shape of the two spans of the bridge as calculated by the algorithm, for different times and relative to the line abutment – second column. The observed 'banana' effect is due to shrinkage of the concrete of the new exterior overhang. This effect stabilizes to a value of 5 mm of horizontal lateral displacement after 1 month.

During a load test, after the end of construction work, the vertical displacement of the bridge was also monitored using the same long-gauge fibre-optic sensors. The diagram of vertical displacements obtained by double integration with assumed boundary conditions that the displacement of the abutment and displacement of the second columns are equal to zero is shown in Figure 4.34. Comparison with the results obtained using traditional invar dial gauges is presented in the same figure. The error of the algorithm is estimated from the deviation from a flat surface of the section deformations. The diagram of vertical displacements obtained using parallel topology and double integration retrieves, within the error interval, the position of the first column (not entered as a boundary condition) and matches the vertical displacement measured with the dial gauges.

4.4 Crossed Topology

4.4.1 Basic Notions on Crossed Topology: Planar Case

A crossed topology (Inaudi and Glišić, 2002b) consists of two crossed sensors installed with a predefined angle with respect to the direction of normal strain lines. The aim of this topology is to detect and quantify the average shear strain in the plane of the sensors. By preference (see below), the sensors have the same gauge length, the angles of sensors are identical by value but different by sign, and the sensors cover the same height of the cross-section. In other words, the sensors combined in a crossed topology are by preference symmetrical with respect to the vertical line crossing their middle points, as shown schematically in Figure 4.35.

The algorithm allowing retrieval of average shear strain depends on the angles of both sensors, their gauge lengths, their positions and strain fields in the equipped cell. In addition, the Bernoulli hypothesis is supposed to be valid.

The measured average strains in the cell i at the position of sensors 1 and 2 at time t are expressed as in Equations 4.9a and 4.9b. Similar to the discussion presented in Section 4.3, the shrinkage has the same influence on the measurements of both sensors ($\varepsilon_{i1,\mathrm{sh},t} = \varepsilon_{i2,\mathrm{sh},t}$). Since the sensors are inclined with the same angle and cover the same height of the cell's cross-section, the temperature changes, including the gradient between the bottom and the top

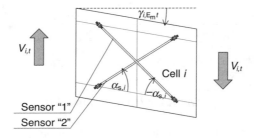

Figure 4.35 Crossed topology (courtesy of SMARTEC).

of the cell, have the same influence on the measurements of both sensors; that is, $\varepsilon_{i1,\text{T},t} = \varepsilon_{i2,\text{T},t}$. Taking into account the discussion above, the measured average shear strain for the crossed topology shown in Figure 4.35 is calculated as follows:

$$\gamma_{i,\text{m},t} = \frac{\varepsilon_{i1,\text{m},t} - \varepsilon_{i2,\text{m},t}}{2\sin\alpha_{\text{s},i}\cos\alpha_{\text{s},i}} = \frac{\varepsilon_{i1,\text{E}_\text{m}+\varphi_\text{m},t} - \varepsilon_{i2,\text{E}_\text{m}+\varphi_\text{m},t}}{2\sin\alpha_{\text{s},i}\cos\alpha_{\text{s},i}} = \gamma_{i,\text{E}_\text{m}+\varphi_\text{m},t} \qquad (4.36)$$

The elastic average shear strain is determined from Equation 4.36 as presented by

$$\gamma_{i,\text{E}_\text{m},t} = \frac{\gamma_{i,\text{E}_\text{m}+\varphi_\text{m},t}}{1 + K_{i,\varphi_\text{m},\text{equ}} f_{i,\varphi_\text{m},\text{equ},t}} = \frac{\varepsilon_{i1,\text{m},t} - \varepsilon_{i2,\text{m},t}}{2\sin\alpha_{\text{s},i}\cos\alpha_{\text{s},i}(1 + K_{i,\varphi_\text{m},\text{equ}} f_{i,\varphi_\text{m},\text{equ},t})} \qquad (4.37)$$

The shear force commonly occurs in combination with a normal (axial) force and bending moments. Thus, the elastic strain measured by sensors can be presented as follows:

$$\varepsilon_{i1,\text{E}_\text{m},t} = \varepsilon_{i1,N,t} + \varepsilon_{i1,M,t} + \varepsilon_{i1,V,t} \qquad (4.38\text{a})$$

$$\varepsilon_{i2,\text{E}_\text{m},t} = \varepsilon_{i2,N,t} + \varepsilon_{i2,M,t} + \varepsilon_{i2,V,t} \qquad (4.38\text{b})$$

For the crossed topology presented in Figure 4.35, the influence of the normal force and bending moments is the same for both sensors, similar to the influence of shrinkage and temperature discussed above; that is, $\varepsilon_{i1,N,t} = \varepsilon_{i2,N,t}$ and $\varepsilon_{i1,M,t} = \varepsilon_{i2,M,t}$. Consequently, the strain components generated by the normal force and bending moment are annulled in Equations 4.36 and 4.37 and these expressions are valid even if the shear force occurs in combination with a normal force and bending moment.

The equivalent creep coefficient and equivalent creep function are to be determined as presented in Sections 3.1.4 and 3.3.3. The relation between the shear force and elastic shear strain is given for linear elastic relation as follows:

$$V_{i,t} = G_{i,t} k_{i,t}^\gamma I_{i,t} \gamma_{i,\text{E}_\text{m},t} = G_{i,t} k_{i,t}^\gamma I_{i,t} \frac{\varepsilon_{i1,\text{m},t} - \varepsilon_{i2,\text{m},t}}{2\sin\alpha_{\text{s},i}\cos\alpha_{\text{s},i}(1 + K_{i,\varphi_\text{m},\text{equ}} f_{i,\varphi_\text{m},\text{equ},t})} \qquad (4.39)$$

where $G_{i,t} = E_{i,t}[2(1 + \nu_{i,t})]^{-1}$ is the shear modulus in cell i at time t, determined from Young's modulus $E_{i,t}$ and Poisson's coefficient $\nu_{i,t}$; $k_{i,t}^\gamma$ is a coefficient that depends on the cross-sectional properties of cell i at time t, determined based on the theory of the *Strength of Materials* (Brčić, 1989).

For inhomogeneous materials, and in particular for concrete, the determination of the coefficient $k_{i,t}^\gamma$ may be very difficult, since the cross-sectional properties depend on the shear reinforcement orientation and the crack orientation and depth.

In general, it is possible to build a crossed topology using sensors with different angles, different positions or different gauge lengths, but in these cases the determination of average shear strain becomes more complex and Equations 4.36, 4.37 and 4.39 are not applicable.

4.4.2 Basic Notions on Crossed Topology: Spatial Case

For beams loaded in one plane, the shear strain is a consequence of the shear force only. In the three-dimensional case, the total shear strains γ_{ix} and γ_{iy} are consequences of the shear

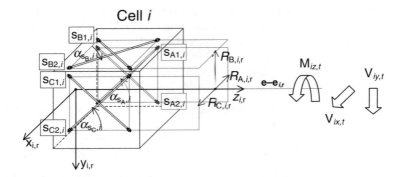

Figure 4.36   Crossed topology, spatial case.

forces V_{ix} and V_{iy} (with respect to the principal axes x_i and y_i) and the torsion moment M_z (see Figure 4.36). However, it is of interest to determine the parameters that are their direct consequences:

$\gamma^*_{ix,m,t}$ is the average shear strain component generated only by shear force $V_{ix,t}$ at time t
$\gamma^*_{iy,m,t}$ is the average shear strain component generated only by shear force $V_{iy,t}$ at time t
$\theta_{i,m,t}$ is the torsion at time t.

Therefore, in the general case, the cell is to be equipped with three crossed topologies. The shear strain distribution in the cell subject to both shear forces and torsion depends on the cross-sectional properties. The shear strain distribution is not the same for full, open thin-walled and closed thin-walled cross-sections. Therefore, the position of the three crossed topologies is to be determined depending on the type of cross-section. For the full cross-section it is recommended to install the crossed topologies in planes parallel to the principal axes, as shown in Figure 4.36.

Let the three crossed topologies be denoted by 'A', 'B' and 'C' and their distances from the reference centre of gravity by $R_{A,i,r}$, $R_{B,i,r}$ and $R_{C,i,r}$, the sensors belonging to them by $s_{A1,i}$ and $s_{A2,i}$, $s_{B1,i}$ and $s_{B2,i}$, and $s_{C1,i}$ and $s_{C2,i}$, and corresponding angles with $\alpha_{s_A,i}$, $\alpha_{s_B,i}$ and $\alpha_{s_C,i}$ (see Figure 4.36). Then each topology measures the average shear strain as follows (see Equation 4.36):

$$\gamma_{iQ,m,t} = \gamma_{iQ,E_m+\varphi_m,t} = \frac{\varepsilon_{iQ1,m,t} - \varepsilon_{iQ2,m,t}}{2\sin\alpha_{s_Q,i}\cos\alpha_{s_Q,i}} \tag{4.40}$$

where $Q \in \{A, B, C\}$.
The influence of torsion on each topology is given by:

$$\gamma_{iQ,m,t}(M_{iz,t}) = \gamma_{iQ,E_m+\varphi_m,t} = \theta_{i,m,t}R_{Q,i,r} = \theta_{i,E_m+\varphi_m,t}R_{Q,i,r} \tag{4.41}$$

where $Q \in \{A, B, C\}$, as in Equation 4.40.
In Equation 4.41 it is important to pay attention to the sign. For example, consider Figure 4.36: the torsion is positive and its influence to topology 'C' is positive, but its influence to topology 'A' is negative.

In general, each topology is influenced by each shear force and torsion moment, but for the example shown in Figure 4.36 the shear force $V_{ix,t}$ does not influence topologies 'A' and 'C' and the shear force $V_{iy,t}$ does not influence topology 'B'. Moreover, the influence of shear force $V_{iy,t}$ is equal for topologies 'A' and 'C'. Finally, the following system of equations is valid:

$$\left.\begin{array}{l} \gamma^*_{iy,\mathrm{m},t} - \theta_{i,\mathrm{m},t} R_{\mathrm{A},i,\mathrm{r}} = \gamma_{i\mathrm{A},\mathrm{m},t} \\[4pt] \gamma^*_{ix,\mathrm{m},t} + \theta_{i,\mathrm{m},t} R_{\mathrm{B},i,\mathrm{r}} = \gamma_{i\mathrm{B},\mathrm{m},t} \\[4pt] \gamma^*_{iy,\mathrm{m},t} + \theta_{i,\mathrm{m},t} R_{\mathrm{C},i,\mathrm{r}} = \gamma_{i\mathrm{C},\mathrm{m},t} \end{array}\right\} \tag{4.42}$$

and the following expressions for unknown shear deformation parameters are obtained:

$$\gamma^*_{ix,\mathrm{m},t} = \gamma^*_{ix,\mathrm{E}_\mathrm{m}+\varphi_\mathrm{m},t} = \gamma_{i\mathrm{B},\mathrm{m},t} + R_{\mathrm{B},i,\mathrm{r}} \frac{\gamma_{i\mathrm{A},\mathrm{m},t} - \gamma_{i\mathrm{C},\mathrm{m},t}}{R_{\mathrm{C},i,\mathrm{r}} + R_{\mathrm{A},i,\mathrm{r}}} \tag{4.43a}$$

$$\gamma^*_{iy,\mathrm{m},t} = \gamma^*_{iy,\mathrm{E}_\mathrm{m}+\varphi_\mathrm{m},t} = \frac{\gamma_{i\mathrm{A},\mathrm{m},t} R_{\mathrm{C},i,\mathrm{r}} + \gamma_{i\mathrm{C},\mathrm{m},t} R_{\mathrm{A},i,\mathrm{r}}}{R_{\mathrm{C},i,\mathrm{r}} + R_{\mathrm{A},i,\mathrm{r}}} \tag{4.43b}$$

$$\theta_{i,\mathrm{m},t} = \theta_{i,\mathrm{E}_\mathrm{m}+\varphi_\mathrm{m},t} = \frac{-\gamma_{i\mathrm{A},\mathrm{m},t} + \gamma_{i\mathrm{C},\mathrm{m},t}}{R_{\mathrm{C},i,\mathrm{r}} + R_{\mathrm{A},i,\mathrm{r}}} \tag{4.43c}$$

A detailed analysis of beams with thin-walled cross-sections is very complex, for which advanced theory concerning the strength of materials has to be employed. That is why only the basic principles are presented here.

In thin-walled beams, the shear forces V_{ix} and V_{iy} generate the shear strain with respect to the wall's middle line; consequently, the crossed topologies are to be installed in the plane parallel to the elastic line of the beam and passing through the wall's middle line. Typical examples are presented schematically in Figure 4.37.

For beams with thin walled cross-sections the following system of equations is valid:

$$\left.\begin{array}{l} K_{x\mathrm{A},t}\gamma^*_{ix,\mathrm{m},t} + K_{y\mathrm{A},t}\gamma^*_{iy,\mathrm{m},t} + \text{or} - K_{\theta\mathrm{A},t}\theta_{i,\mathrm{m},t} R_{\mathrm{A},i,\mathrm{r}} = \gamma_{i\mathrm{A},\mathrm{m},t} \\[4pt] K_{x\mathrm{B},t}\gamma^*_{ix,\mathrm{m},t} + K_{y\mathrm{B},t}\gamma^*_{iy,\mathrm{m},t} + \text{or} - K_{\theta\mathrm{B},t}\theta_{i,\mathrm{m},t} R_{\mathrm{B},i,\mathrm{r}} = \gamma_{i\mathrm{B},\mathrm{m},t} \\[4pt] K_{x\mathrm{C},t}\gamma^*_{ix,\mathrm{m},t} + K_{y\mathrm{C},t}\gamma^*_{iy,\mathrm{m},t} + \text{or} - K_{\theta\mathrm{C},t}\theta_{i,\mathrm{m},t} R_{\mathrm{C},i,\mathrm{r}} = \gamma_{i\mathrm{C},\mathrm{m},t} \end{array}\right\} \tag{4.44}$$

where the coefficient K depends on the cross-sectional properties at time t (thickness of the wall at location of topology and exact position of the topology with respect to the middle line), and '$+$ or $-$' is a reminder to pay attention to determination of the sign of the torsion.

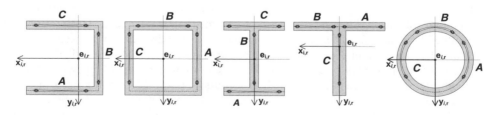

Figure 4.37 Example of recommended cross-sectional positions of crossed topologies, thin-walled cross-sections.

The relations between the shear forces and elastic shear strains are given for linear elastic relation as follows:

$$V_{ix,t} = G_{i,t} k_{ix,t}^{\gamma} I_{iy,t} \gamma_{ix,\mathrm{E_m},t}^* = G_{i,t} k_{ix,t}^{\gamma} I_{iy,t} \frac{\gamma_{ix,\mathrm{E_m}+\varphi_\mathrm{m},t}^*}{1 + K_{i,\varphi_\mathrm{m},\mathrm{equ}} f_{i,\varphi_\mathrm{m},\mathrm{equ},t}}$$

$$= G_{i,t} k_{ix,t}^{\gamma} I_{iy,t} \frac{\gamma_{ix,\mathrm{m},t}^*}{1 + K_{i,\varphi_\mathrm{m},\mathrm{equ}} f_{i\,\varphi_\mathrm{m},\mathrm{equ},t}} \tag{4.45a}$$

$$V_{iy,t} = G_{i,t} k_{iy,t}^{\gamma} I_{ix,t} \gamma_{iy,\mathrm{E_m},t}^* = G_{i,t} k_{iy,t}^{\gamma} I_{ix,t} \frac{\gamma_{iy,\mathrm{E_m}+\varphi_\mathrm{m},t}^*}{1 + K_{i,\varphi_\mathrm{m},\mathrm{equ}} f_{i,\varphi_\mathrm{m},\mathrm{equ},t}}$$

$$= G_{i,t} k_{iy,t}^{\gamma} I_{ix,t} \frac{\gamma_{iy,\mathrm{m},t}^*}{1 + K_{i,\varphi_\mathrm{m},\mathrm{equ}} f_{i,\varphi_\mathrm{m},\mathrm{equ},t}} \tag{4.45b}$$

where coefficients $k_{ix,t}^{\gamma}$ and $k_{iy,t}^{\gamma}$ depend on the cross-sectional properties of cell i at time t, determined based on the theory of the strength of materials (see comments after Equation 4.39 in Section 4.4.1).

The torsion moment is determined using

$$M_{iz,t} = G_{i,t} I_{iz,t} \theta_{i,\mathrm{E_m},t} = G_{i,t} I_{iz,t} \frac{\theta_{i,\mathrm{E_m}+\varphi_\mathrm{m},t}}{1 + K_{i,\varphi_\mathrm{m},\mathrm{equ}} f_{i,\varphi_\mathrm{m},\mathrm{equ},t}}$$

$$= G_{i,t} I_{iz,t} \frac{\theta_{i,\mathrm{m},t}}{1 + K_{i,\varphi_\mathrm{m},\mathrm{equ}} f_{i,\varphi_\mathrm{m},\mathrm{equ},t}} \tag{4.46}$$

where $I_{iz,t}$ is the torsion moment of inertia of cell i at time t, determined based on the theory of the strength of materials.

A crossed topology is mainly used as a complement to a parallel topology, but it can also be used in an independent way for monitoring stiffenings, walls, and so on. An On-site application of crossed topology is presented next.

4.4.3 Example of a Crossed Topology Application

The interstate highway bridge I-10, close to Las Cruces, New Mexico, consists of five spans, and each span of six lines of prefabricated prestressed skewed girders, with a cast-on-site superstructure (Idriss et al., 2006). The girders' cross-sections have a 'U' shape with wings. The girders were cast over prestressed strands in a prefabrication plant and then steam cured for a minimum of 2 days. After the cure was finished, the strands were cut and the prestressing force applied to the girders. The girders were stored for 2 months and then transported to the site, where they were put in place and covered by a cast-on-site superstructure slab.

The aim of monitoring was to increase knowledge of the structural behaviour, to verify hypotheses and to ensure safety (Hughs and Idriss, 2006). The monitoring parameters were average strain, average shear strain, average curvature, deformed shape and prestress losses (Hughs et al., 2005). For this purpose it was decided to equip all the girders of the fifth span, laying on the abutment, with sensors in different configurations. First, two girders were fully equipped with sensors. Three non-coplanar parallel sensors were installed in five

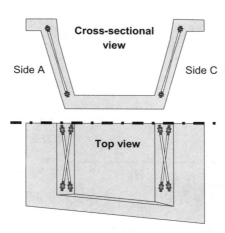

Figure 4.38 Position of crossed topologies in skewed girder cross-section (courtesy of SMARTEC and New Mexico State University).

cross-sections over girders in order to monitor both the horizontal and the vertical deformed shape. Thermocouples were installed in three cross-sections in order to measure the temperature variations necessary to separate thermally generated strain from structural strain. Other girders were equipped with fewer sensors that are used as double-check and redundancy. Two crossed topologies, necessary to evaluate the vertical average shear strain and torsion, were installed in the cross-sections closest to the abutment of each girder. The position of the crossed topologies is shown in Figure 4.38.

The sensors were embedded in the girders during the fabrication. Thus, they provided for full-life measurements of girders, including the very early age and prestressing. The system is fully centralized and the measurements are performed automatically from a control room built on-site. The sensors during installation are shown in Figure 4.39.

The length of the sensors was 2 m and the inclination angle, imposed by the height of the cross-section, was 35.7°. Measurements started immediately after the pouring. In this way the

Figure 4.39 Long-gauge sensors installed on passive rebars before pouring (courtesy of SMARTEC).

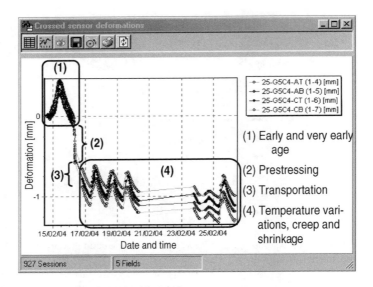

Figure 4.40   Typical results obtained on the I-10 bridge by sensors forming a crossed topology (courtesy of SMARTEC, data by New Mexico State University).

early and very early age deformation were recorded during the first 3 days. The deformation is later recorded during the prestressing phase, after each strand was cut. Thus, the real initial strain state of the girders was recorded. After the prestressing, the girders were transported to a storage area where they were left until transportation to the site and pouring of the deck. Typical results of monitoring obtained during the first 12 days are presented in Figure 4.40.

Prestressing created bending of the girder (camber of ~85 mm in the middle of the span); consequently, the dead load of the girder was 'activated' and it changed the static system to a simple beam supported at the extremities. The reactions at the extremities introduced a vertical shear force and shear strain; also, since the girder was skewed, torsion could occur. The shear strain evolutions on sides A and C, calculated using Equation 4.36, are shown in Figure 4.41.

The shear strain due to the vertical force is calculated using Equation 4.43b. Taking into account that the cross-section is symmetrical with respect to the vertical axis and that the topologies' positions are also symmetrical with respect to the same axis, the following expression for shear strain due to vertical force is obtained:

$$\gamma^*_{y,m,t} = \gamma^*_{y,E_m+\varphi_m,t} = \frac{\gamma_{A,m,t}R_r + \gamma_{C,m,t}R_r}{2R_r} \tag{4.47}$$

where $R_{A,r} = R_{C,r} = R_r$ due to symmetry.

The shear strain due to torsion is determined using Equation 4.43c, as follows:

$$\gamma^*_{tor,m,t} = R_r\theta_{m,t} = R_r\theta_{E_m+\varphi_m,t} = \frac{-\gamma_{A,m,t} + \gamma_{C,m,t}}{2} \tag{4.48}$$

The shear strain due to the vertical force is important and its evolution is noticed in Figure 4.41, whereas the torsion is very stable and small and can practically be neglected.

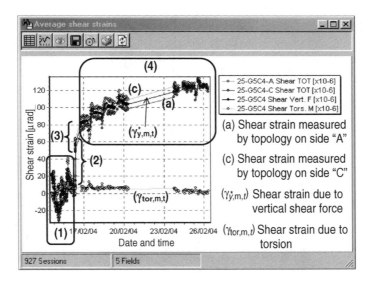

Figure 4.41 Shear strain evolutions determined from crossed topologies (courtesy of SMARTEC, data by New Mexico State University).

The crossed topologies allowed collection of very important information related to the initial state and condition of prestressed girders, and provided input data for numerical modelling and structural condition assessment in the long term.

4.5 Triangular Topology

4.5.1 Basic Notions on Triangular Topology

A triangular topology is used in cells exposed to all types of loads where planar deformation can be described by relative displacement of points. Typical examples are masonry arches, open cracks and joints, and walls, often making a part of historical and heritage structures, but also plates and shells with normal forces only. Contrary to the topologies presented in previous sections, where the primary monitored parameter was strain or shear strain, the primary parameter monitored in the case of a triangular topology is relative displacement.

The triangular topology consists of two sensors that have the one common point and their anchoring points form a triangle, as shown in Figure 4.42.

The points A_{i1}, A_{i2} and B_i define cell i, and sensors i_1 and i_2 are installed respectively between points A_{i1} and B_i and between points A_{i2} and B_i, as shown in Figure 4.42. In the figure, xOy is the local coordinate system related to the monitored structure, and the horizontal relative displacements are denoted by u and the vertical displacements by v. The aim of the triangular topology is to determine the relative displacements of point B_i at time t ($u_{iB,t}$ and $v_{iB,t}$) with respect to a reference position having provided that the relative displacements of points A_{i1} and A_{i2} are known, as well as the measurements of the sensors i_1 and i_2 ($m_{s1,i,t}$ and $m_{s21,i,t}$).

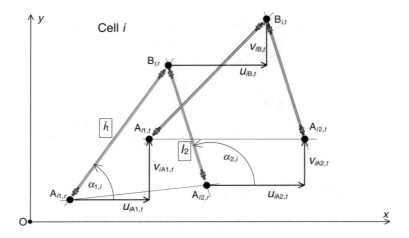

Figure 4.42   Triangular topology schematic.

Owing to the physical dimensions, it might be impossible to install the sensors exactly from point to point; thus, in the general case, the sensors are shorter than the distance between the points (see Figure 4.42). In order to determine the relative displacements between the points, the sensor measurements are first extrapolated from the gauge length to the real distance between the points as given in the following expressions:

$$m^*_{s1,i,t} = m_{s1,i,t} \frac{L_{A_{i1,r}-B_{i,r}}}{L_{s1,i}} = \varepsilon_{i1,m,t} L_{A_{i1,r}-B_{i,r}} \tag{4.49a}$$

$$m^*_{s2,i,t} = m_{s2,i,t} \frac{L_{A_{i2,r}-B_{i,r}}}{L_{s2,i}} = \varepsilon_{i2,m,t} L_{A_{i2,r}-B_{i,r}} \tag{4.49b}$$

where $m_{sk,i,t} = \Delta L_{sk,i,t}$ is the relative displacement between sensor anchoring points measured by sensor k (see Figure 3.8 and Equation 3.23).

The relative displacements of point B_i at time t ($u_{iB,t}$ and $v_{iB,t}$) are determined using

$$u_{iB,t} = -m^*_{s1,i,t} \frac{\sin \alpha_{2,i}}{\sin(\alpha_{1,i} - \alpha_{2,i})} + m^*_{s2,i,t} \frac{\sin \alpha_{1,i}}{\sin(\alpha_{1,i} - \alpha_{2,i})}$$

$$+ \frac{-u_{iA1,t} \cos \alpha_{1,i} \sin \alpha_{2,i} + u_{iA2,t} \sin \alpha_{1,i} \cos \alpha_{2,i}}{\sin(\alpha_{1,i} - \alpha_{2,i})} + \frac{(-v_{iA1} + v_{iA2}) \sin \alpha_{1,i} \sin \alpha_{2,i}}{\sin(\alpha_{1,i} - \alpha_{2,i})} \tag{4.50a}$$

$$v_{iB,t} = m^*_{s1,i,t} \frac{\cos \alpha_{2,i}}{\sin(\alpha_{1,i} - \alpha_{2,i})} - m^*_{s2,i,t} \frac{\cos \alpha_{1,i}}{\sin(\alpha_{1,i} - \alpha_{2,i})} + \frac{(u_{iA1} - u_{iA2}) \cos \alpha_{1,i} \cos \alpha_{2,i}}{\sin(\alpha_{1,i} - \alpha_{2,i})}$$

$$+ \frac{v_{iA1,t} \sin \alpha_{1,i} \cos \alpha_{2,i} - v_{iA2,t} \cos \alpha_{1,i} \sin \alpha_{2,i}}{\sin(\alpha_{1,i} - \alpha_{2,i})} \tag{4.50b}$$

The use of a triangular topology in monitoring small arches and large surfaces is presented next.

4.5.2 Scattered and Spread Triangular Topologies

A triangular topology is, in general, used in a 'scattered' or 'spread' manner. A scattered triangular topology is applied when the structure consists of a number of elements or monitoring positions of interest that are not close to each other. Typical examples are roofs supported by small arches or vaults, as shown in Figure 4.43.

In Figure 4.43, a third, horizontal sensor is added to each triangular topology. The local coordinate system for each topology is different, and its origin O_i coincides with point A_{i1} ($O_i \equiv A_{i1}$). Then, each topology measures the relative displacement between the bases and the tops of each arch (i.e. settlements of the arches, which may be the consequence of different issues related to horizontal and vertical relative displacements of an arch base (relative settlement or relative inclination of supporting columns), or excessive vertical loads in the arches). Since the sensor measures relative displacement only in the direction of the sensor, the displacement in the perpendicular direction is determined indirectly from the relative rotation of the sensor, measured using an inclinometer (tilt-meter). In order to determine the relative displacements between the points, the following known displacements of points A_{i1} and A_{i2} are introduced in Equations 4.50a and 4.50b:

$$u_{iA1,t} = v_{iA1,t} = 0 \qquad (4.51a)$$

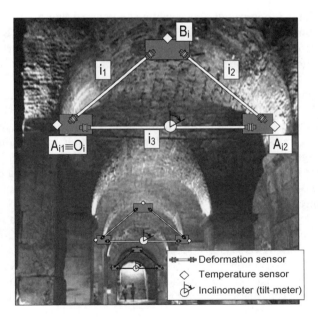

Figure 4.43 Application example of a scattered triangular topology (courtesy of SMARTEC).

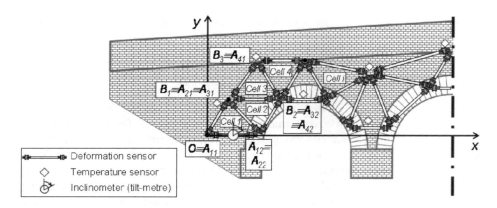

Figure 4.44   Application example of a spread triangular topology (courtesy of SMARTEC).

$$u_{iA2,t} = m^*_{s3,i,t} \qquad\qquad (4.51b)$$

$$v_{iA2,t} = \omega^*_{iA1,t} L_{A_{i1,r}-A_{i2,r}} \qquad\qquad (4.51c)$$

where $\omega^*_{iA1,t}$ is the rotation of the axis A_{i1}–A_{i2}, measured with the inclinometer (tilt-meter).

The positions of inclinometers in Figure 4.43 are only presented schematically. Actually, special nondeformable supports with free rotation at points A_{i1} and A_{i2} is to be built in order to hold the inclinometers. In such a manner, the rotation of the support is the same as the rotation of the sensor i_3. The results of monitoring obtained from the different topologies are analysed, taking into account the static system of the structure as a whole and the boundary conditions.

A spread triangular topology is applied in cases where the planar displacements are of interest; for example, for plates, shells and walls subject to normal (axial) forces or for large masonry structures where the use of parallel and crossed topologies is not efficient. An example of the use of a spread triangular topology for masonry structure monitoring is presented schematically in Figure 4.44. The aim is to retrieve the deformed shape of the bridge. Point A_{11} is selected to be the origin of the local coordinate system ($xA_{11}y \equiv xOy$), the sensor between the points A_{11} and A_{12} is horizontal and its rotation is monitored using an inclinometer, similar to the example presented above.

The point $O \equiv A_{11}$ (see notation in Figures 4.42 and 4.44) is taken as a reference and the relative displacements of points $A_{12} \equiv A_{22}$ and $B_1 \equiv A_{21} \equiv A_{31}$ are calculated using Equations 4.51a–4.51c and Equations 4.50a and 4.50b respectively.

The algorithm presented in Equations 4.50a and 4.50b is then successively repeated for other cells until the relative displacement of all the points with respect to the reference coordinate system is determined. The calculated distribution of horizontal and vertical relative displacements will provide for determination of the deformed shape of the bridge. It is possible to detect and quantify undesired events based on the deformed shape, such as permanent deflection of the bridge, differential settlements of supports and columns, settlements and openings of the arches, overloading and so on.

4.5.3 Monitoring of Planar Relative Movements Between Two Blocks

Monitoring of planar relative movements between two blocks (e.g. relative movements of blocks separated by a crack or joint) is frequently of interest in practice. The planar relative movement consists, in the general case, of two relative translations (along the x- and y-axes) of one point (B_1) and one relative rotation between the blocks ψ_t. Supposing that the deformation of the blocks is much smaller than their relative displacements, then the relative planar movement is monitored using two triangular topologies, as shown schematically in Figure 4.45.

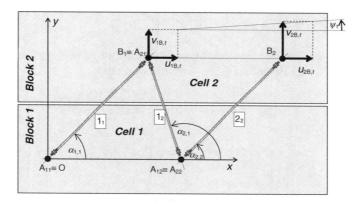

Figure 4.45   Monitoring of relative movement between two blocks using two triangular topologies.

Let points $O_1 \equiv A_{11}$ and $A_{12} \equiv A_{22}$ belong to block 1, let points $B_1 \equiv A_{21}$ and B_2 belong to block 2 and let segments $A_{11}A_{12}$ and B_1B_2 be parallel. Since the deformation of the blocks is much smaller than the relative displacements between them, the following expressions are valid:

$$u_{1A1,t} = v_{1A1,t} = u_{1A2,t} = v_{1A2,t} = 0 \tag{4.52a}$$

$$A_{11}A_{12}\|B_1B_2 \Rightarrow \angle(O_1x)(B_1B_2) = 0 \tag{4.52b}$$

$$m^*_{sj,i,t} \approx m_{sj,i,t} \tag{4.52c}$$

The relative displacement of point B_1 (i.e. the relative translations of a block along axes x and y) is determined using the conditions in Equations 4.52a–4.52c inserted in Equations 4.50a and 4.50b:

$$u_{1B,t} = -m_{s1,1,t}\frac{\sin\alpha_{2,1}}{\sin(\alpha_{1,1} - \alpha_{2,1})} + m_{s2,1,t}\frac{\sin\alpha_{1,1}}{\sin(\alpha_{1,1} - \alpha_{2,1})} \tag{4.53a}$$

$$v_{1B,t} = m_{s1,1,t} \frac{\cos \alpha_{2,1}}{\sin(\alpha_{1,1} - \alpha_{2,1})} - m_{s2,1,t} \frac{\cos \alpha_{1,1}}{\sin(\alpha_{1,1} - \alpha_{2,1})} \tag{4.53b}$$

The relative displacement of point B_2 is determined using the conditions in Equations 4.52a–4.52c, 4.53a and 4.53b inserted in Equations 4.50a and 4.50b:

$$u_{2B,t} = u_{1B,t} \tag{4.54a}$$

$$v_{2B,t} = m_{s2,2,t} \frac{1}{\sin \alpha_{2,2}} - u_{1B,t} \frac{\cos \alpha_{2,2}}{\sin \alpha_{2,2}} \tag{4.54b}$$

And finally, the relative rotation between the blocks is calculated thus:

$$\psi_t = \frac{v_{2B,t} - v_{1B,t}}{L_{B_1-B_2}} \tag{4.55}$$

Equations 4.53a–4.55 can be simplified by suitable selection of angles α. Let $\alpha_{1,1} = \alpha_{2,2} = 90°$ and $\alpha_{2,1} = 135°$, then Equations 4.53a–4.55 transform to

$$u_{1B,t} = u_{2B,t} = m_{s1,1,t} - m_{s2,1,t} \frac{\sqrt{2}}{2} \tag{4.56a}$$

$$v_{1B,t} = m_{s1,1,t} \tag{4.56b}$$

$$v_{2B,t} = m_{s2,2,t} \tag{4.56c}$$

$$\psi_t = \frac{m_{s2,2,t} - m_{s1,1,t}}{L_{B_1-B_2}} \tag{4.56d}$$

4.5.4 Example of a Triangular Topology Application

Vienna, the capital of Austria, is supplied with water directly from the Alps. The second Vienna water supply tunnel was built in 1900. It was made of concrete, with a diameter of more than 2 m and a total length of 170 km, approximately. A number of cracks, which cause water losses, appeared in some of the sections a long time ago. They were sealed with an appropriate treatment, but significant leakage was still present.

In order to increase our knowledge and assess the condition of the water supply tunnel, it was decided to perform long-term monitoring. Besides convergence and strain monitoring, crack monitoring was of particular interest. Three cracks along 20 m of tunnel were equipped with triangular topologies, each consisting of only two sensors with angles of 45° and 135°. These topologies provided for monitoring longitudinal movement (crack opening) and the shear movement of the cracks. A third sensor (see Section 4.5.3) was not necessary since no relative rotation occurs in the crack. The position of the sensors in the cross-section is shown in Figure 4.46.

Figure 4.46   Triangular topology in the water supply tunnel (courtesy of SMARTEC and RISS).

Monitoring was performed over several years and during the filling and emptying of the pipe. The data are used to plan additional repair interventions.

The measurements show that the cracks tend to open slowly as time passes. This small tendency overlaps with the larger deformations due to the filling of the pipe. Shearing effects are also observed.

5

Finite Element Structural Health Monitoring Strategies and Application Examples

5.1 Introduction

The basis of the FE-SHM (Finite Element-structural Health Monitoring) concept was presented in Chapter 4. The particularity of the method is the use of deformation fibre-optic sensors combined in different topologies. The structure is divided in to cells, each cell is equipped with a sensor topology corresponding to the expected strain field in the cell, and then the results obtained from each cell are linked in order to retrieve the global structural behaviour.

Sensor topologies, combined with temperature sensors and inclinometers, allow monitoring of primary parameters such as:

1. Average strain
2. Average shear strain
3. Average curvature
4. Relative displacements
5. Inclination (rotation)
6. Temperature
7. Temperature gradients.

The following secondary parameters, related to deformation, can indirectly be determined from the primary parameters using appropriate FE-SHM algorithms, as presented in Chapter 4:

(A) Deformed shape
(B) Displacement distribution

Fibre Optic Methods for Structural Health Monitoring   B. Glišić and D. Inaudi
© 2007 John Wiley & Sons, Ltd

Finally, the structural parameters related to structural malfunctioning can be determined from primary and secondary determined parameters:

(C) Crack detection, distribution and characterization.
(D) Prestress losses
(E) Differential settlements of foundations
(F) Damage in structural members
(G) Other.

In order to achieve the aims of monitoring it is necessary to employ a good monitoring strategy. FE-SHM offers a good basis for making decisions concerning the monitoring strategy. A good monitoring strategy can provide excellent results with a relatively limited budget. The selection of the sensor network to be used for monitoring depends on the type of structure, the expected loads, possible sources of damage and degradation phenomena expected to occur during the structure's lifespan.

Different structures are subject to different service conditions, and the diversity of monitoring cases is so large that the amount of possible monitoring solutions is practically innumerable; consequently, a universal solution that will cover all the possible cases does not exist. However, FE-SHM proposes typical solutions for the most common types of structure, such as piles, buildings, bridges, dams, tunnels and heritage structures, and the most frequently monitored parameters, such as those listed above.

The case studies of the most frequent types of structure are presented schematically in this chapter and illustrated with on-site examples, when available. General principles are presented without entering in to detailed numerical modelling. In addition to classic civil engineering structures, solutions for some structures from oil and gas industry, such as pipelines, are also given. The presented case studies are based on engineering studies prepared for real applications. In order to facilitate the reading and comprehension of schemas and drawings, the following legend of symbols was created:

We attempt to be as complete as possible, but lists of the monitored parameters and methods presented should not be considered as exhaustive. For any other monitoring case, it is recommended to consult all the sections of this chapter and to combine the presented solutions in order to define the best monitoring strategy.

5.2 Monitoring of Pile Foundations

5.2.1 Monitoring the Pile

A part of the monitoring strategy for piles is presented in Sections 4.2 and 4.3. The principles developed in those sections are completed here.

Piles subject only to axial load (compression or pullout) are monitored with an enchained simple topology. The following recommendations are given:

1. In order to determine the normal (axial) force in the head and in the tip of the pile with better accuracy, it is recommended to equip the first (top) and the last (bottom) cell with

sensors having a gauge length between 1 and 2 m; head sensors are installed just below the foundation slab.

2. For piles designed to transmit the loads to the soil only by friction, it is recommended to equip the head of the pile with a parallel topology (for biaxial bending in general) in order to detect eccentricity of the load.

3. For piles designed to transmit the loads to the lower supporting layer of soil, both the head and the tip are to be equipped with parallel topologies in order to detect eccentricity of both the load and bottom force.

4. The best results of monitoring are achieved if an enchained simple topology covers all the length of the pile; in this case, the following parameters are monitored:

   a. Eccentricity
   b. Load (normal force in the head of the pile), if the Young's modulus of the pile and cross-sectional properties are known
   c. Young's modulus of the pile (if the load and cross-sectional properties of the pile are known)
   d. Distribution of average strain along the pile
   e. Deformation of the pile (deformed shape and vertical displacement distribution, as long as no vertical movements in the tip are detected)
   f. Force distribution along the pile (if Young's modulus is known)
   g. Strain when cracks occur (for piles exposed to pullout)
   h. Damage to the pile (detection and localization, for piles exposed to pullout)
   i. Properties of the soil (capacity of different layers)
   j. Forces in the soil (frictional stresses distribution in the soil)
   k. Failure mode (slipping of the pile on soil or cracking on the pile)
   l. Failure by buckling can be detected with an enchained simple topology, but if there is a risk of buckling, then it is recommended to equip the pile with parallel topologies in order to avoid ambiguity in data analysis
   m. Ultimate load capacity.

5. If the budget for monitoring is reduced, at least the head and the tip of the pile are to be equipped with sensors; in this case, points a, b, c, f, j (only the average force distribution and average friction along the pile, and not the full distribution), k (ambiguity can occur: due to buckling the head and the tip deformation will decrease, but for whom analyses the data it is not clear if the deformation decreased due to buckling or to decrease in load).

6. Where the cross-sectional properties of the pile change at some points along the pile (e.g. less rebars or even absence of reinforcement in some parts of the pile) it is recommended to install sensors on both sides of this particular cross-section.

Complete and reduced solutions, as presented in point (5), for axially loaded piles are presented in Figure 5.1.

Piles subject to bending (horizontal force on the head of the pile), with or without simultaneous axial load, and piles subject only to compression force but with high risk of buckling are monitored using a parallel topology, as presented in Figure 4.4. The following recommendations are given for this case:

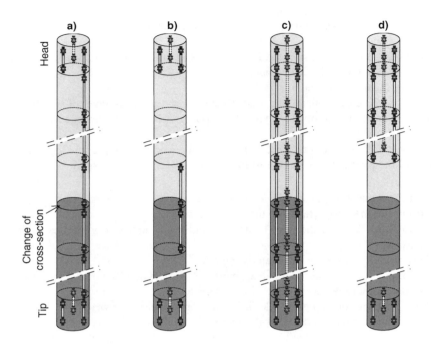

Figure 5.1 Complete and reduced solutions for piles: (a) complete solution for axially loaded pile; (b) reduced solution for axially loaded pile whose cross-sectional properties change; (c) complete solution for pile subject to biaxial bending with or without simultaneous axial loading; (d) reduced solution for pile subject to biaxial bending without axial loading.

1. In general, the horizontal load can have any direction, and the parallel topology for biaxial bending is to be used (three sensors per cell); if the direction of horizontal load is known, then the plane of bending of the pile is known and a parallel topology for uniaxial bending can be used (two sensors per cell) – this decreases the cost of equipment; head sensors are installed just below the foundation slab.
2. The best monitoring results are achieved if parallel topologies cover all the length of the pile; in this case, the following parameters are monitored:
   a. Load (horizontal forces in the head of the pile), if the Young's modulus of the pile and cross-sectional properties are known
   b. Young's modulus of pile (if the load and cross-sectional properties of the pile are known)
   c. Distribution of curvature along the pile
   d. Deformation of pile (deformed shape and horizontal displacement, as long as no horizontal movements or rotations in the tip are detected)
   e. Force and bending moments distribution along the pile (if Young's modulus and cross-sectional properties are known)
   f. Strain when cracks occur
   g. Damage to the pile (detection and localization)
   h. Forces in the soil (horizontal reactions distribution in the soil)
   i. Failure mode (horizontal slipping of the pile on soil or cracking and breaking on the pile)

j. Failure by buckling with localization of buckling area

k. Ultimate load capacity.

3. If the budget for monitoring is reduced, then at least the head, the upper part of the pile and the tip of the pile are to be equipped with sensors; this is justified by the fact that bending in the pile is localized in the upper part and rarely propagates to the bottom of the pile; monitoring parameters presented in the above points a–k are limited to only that part of the pile equipped with sensors.

4. Where the cross-sectional properties of the pile change at some points along the pile (e.g. less rebars or even absence of reinforcing in some parts of the pile) it is recommended to install the sensors on both sides of this particular cross-section; in the case of a reduced budget with the condition that only a horizontal force loads the pile, it is not necessary to equip this section if it is located lower than where the bending influence is expected

5. Sensor positions in a cell can, in general, be arbitrary; but, for reasons of simplification of data analysis, it is recommended to install them either with rotational symmetry with an angle of $120°$ or with symmetry with respect to the main axes, as presented in Figure 5.2.

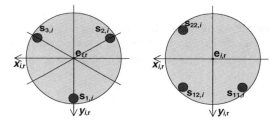

Figure 5.2   Example of recommended cross-sectional positions of sensors in piles subject to biaxial bending.

Complete and reduced solutions, as presented in points (3) and (4), for piles subject to biaxial bending are presented in Figure 5.1.

For all the cases presented in this section it is important to evaluate the necessity of installing redundant sensors. Redundancy will increase the cost of monitoring, but will allow for a more accurate and more reliable measurement system.

5.2.2 Monitoring a Group of Piles

Pile foundations commonly consist of a thick foundation slab cast over a group of piles. The aim of the thick slab is to redistribute the loads coming from the structure to the piles. Since the number of piles below the slab can be very important, monitoring of all the piles would require a significant number of sensors; consequently, the cost of monitoring would be very high. In order to reduce the cost of monitoring it is recommended to monitor only the representative piles belonging to the same foundation; that is, those found below the same foundation slab.

The selection of representative piles to be monitored depends on the type of load applied (vertical and horizontal forces, and bending and torsion moments), the position of the pile with respect to the centre of gravity of the foundation slab and the cross-sectional properties of pile.

Supposing that all the piles have the same cross-sectional properties (the same stiffness), and with the assumption that the foundation slab is very (infinitely) stiff, the vertical and horizontal forces are supposed to generate approximately the same forces in each pile; thus, the minimum number of representative piles is one, and this pile can be selected in an arbitrary location. However, it is recommended to monitor at least three non-coplanar piles in order to better evaluate the load distribution in the pile (detect moment if any), evaluate boundary conditions between the slab and piles and for redundancy reasons.

The more distant the piles are from the slab's centre of gravity, the more they are loaded by the bending and torsion moments; thus, if all the piles have the same cross-sectional properties, the most distant piles should be selected for monitoring. Besides these most loaded piles also some piles with lower load (i.e. closer to the slab's centre of gravity) are to be monitored in order to determine the load distribution in the piles more accurately and for comparison reasons. A minimum three piles is to be monitored, but this number can be increased to six, depending on the boundary conditions between the slab and piles. As in the case of a foundation loaded only with horizontal and vertical forces, it is recommended to monitor more piles than the minimum, with the aim to determine the load distribution in piles with better accuracy and to provide some redundancy.

Finally, where some piles have different cross-sectional properties (different stiffness), or different inclinations, it is recommended to monitor at least one pile of each different type.

The engineer responsible for defining the monitoring strategy for the pile foundation is supposed to analyse the load distribution in the piles and select representative piles. Some (but not all) examples of load cases and the corresponding representative piles are given in Figure 5.3.

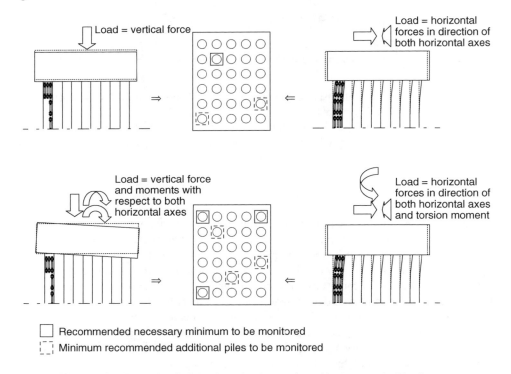

Figure 5.3   Example of piles selected to be monitored in some typical load cases.

5.2.3 *Monitoring of Foundation Slab*

The foundation slab is usually designed and considered as infinitely stiff. In that manner, the loads are distributed to the piles in a linear manner. However, the slab is not, in reality, infinitely stiff: due to heavy reinforcement and the small distance between the piles, the strain distribution in the slab does not follow the rules of linear theory (theory of the first order). The loads in the slab can be determined indirectly from monitoring of the piles . Nevertheless, it is of interest to monitor some parts of the slab directly, notably those points with strong tensional forces.

Tensional stresses are expected to occur in the slab near to the head of axially loaded piles as a consequence of shear forces. In the case of horizontally loaded piles, tension occurs in the horizontal plane. Since the number of piles can be large, it is recommended to select for monitoring the most critical locations in the slab, from the point of view of loads and the stiffness (reinforcing), but also the boundary conditions of the slab in this particular location.

The parameters recommended to be monitored are the shear strain near to head of the axially loaded piles and the horizontal strain near to the heads of piles loaded with a horizontal force. Some examples are presented in Figure 5.4.

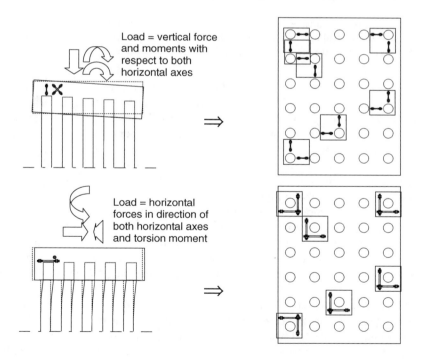

Figure 5.4   Example of slab locations selected to be monitored for some typical load cases.

5.2.4 On-Site Example of Piles Monitoring

An on-site example of piles monitoring was partially presented in Sections 4.2.3 and 4.3.4
(Glišić *et al.*, 2002b). The part concerning the installation of the sensors is presented here.
Particular attention is given to sensor protection, since geotechnical work, in general, presents
very difficult installation conditions.

The sensors were installed to the main rebars using plastic ties. This manner of installation
was fast and safe. In order to avoid the influence of connecting cables to the sensors, they
were attached to a neighbouring rebar. The sensors installed on the rebars and the detail of the
anchor parts of the sensors attached with plastic ties are shown in Figure 5.5.

Figure 5.5 (a) Sensors (indicated with black arrows) installed on rebars and passive cables (indicated
with white arrows) attached to a neighbouring rebar; (b) detail of anchor pieces (black arrows) attached
with plastic ties (white arrows) (courtesy of SMARTEC and RouteAero).

The rebars cage of these piles was too long to be put into the borehole all at once (see
Figure 4.4). Therefore, it was split in to three sections, which were lowered down sequentially
and assembled by welding. The sensors were first installed on each section (indicated with
I–VI in Figure 4.4) and the sensors whose positions corresponded to a welded region were
installed after welding, while lowering the cage (indicated with VII and VIII in Figure 4.4).
In order to prevent damage to the sensors during welding, the sensors installed on rebars were
protected with special fireproof mats. The lowering of the cages and installation of sensors in
the welded regions are shown in Figure 5.6.

Once the cage was completely put in place, the cables were grouped at one location and
protected with a metallic pipe in order to provide for the cables' necessary mechanical protection. Immediately after the installation work was completed, the concrete was poured. The
functioning of the sensors was verified in parallel with the pouring work. The cables just after
the installation and the checking of sensor function during the pouring work are shown in
Figure 5.7.

After 28 days, the upper parts of the piles were cut and prepared for the tests. The metallic
tube provided the necessary protection during the cutting. In spite of very difficult installation
conditions (welding of rebars, lowering of rebar cages, submersion in bentonite and cutting of
upper parts of the pile) the overall survival rate was better than 95 %, proving that fibre-optic
sensors can be applied in challenging conditions.

Figure 5.6   (a) Alignment of rebars after lowering the cage; black arrows indicates sensors protected with fireproof mats. (b) Installation of sensor (indicated with black arrow) after the welding of rebars (courtesy of SMARTEC and RouteAero).

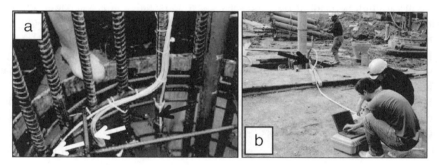

Figure 5.7   (a) The egress point of passive cables (indicated with white arrows) grouped together; the head sensor is indicated with black arrow. (b) Checking the sensors during the pouring of concrete; the passive cables are protected with metallic pipe, indicated with a black arrow (courtesy of SMARTEC and RouteAero).

5.3 Monitoring of Buildings

5.3.1 Monitoring of Building Structural Members

Part of the monitoring strategy for buildings were presented in Sections 3.3 and 4.2. The principles developed in those sections are completed here.

Buildings consist of structural members subject to vertical loads (dead loads and live loads) and horizontal loads (wind and earthquake). These structural members may be clearly distinguishable (vertical loads are supported by slabs, frames and columns and horizontal loads are supported by cores) or combined (the supporting wall can accept both vertical loads from slabs and upper storeys as well as the horizontal loads). Finally, the loads are guided to the foundations. The monitoring of pile foundations was presented in the previous section and will not be discussed further; the focus below is mostly on the monitoring of columns, cores and supporting walls.

The main issue in defining the monitoring strategy for buildings is actually the large amount of structural members that are potential candidates for monitoring. For high-rise buildings, which are actually more subject to monitoring than low-rise buildings, the number of structural members can reach several hundreds or even exceed a thousand. Therefore, an affordable monitoring strategy is to be developed considering a reduced number of sensors. First, the monitoring strategies for structural members are presented and then a monitoring strategy is developed for the building as a whole.

5.3.2 Monitoring of Columns

Building columns can be subject dominantly to normal (axial) forces or to the combined influence of normal forces and bending moments. Only those columns with constant cross-sectional properties are discussed in this section. Where the columns are built as bending free, a simple topology can be used for their monitoring, as presented in Section 3.3.4. The following main principles are recommended:

1. Since the column is bending free, the sensor can be installed at any location in the cross-section; the position in the cross-section can be determined from the following requirements:
   a. If the risk of buckling of column is minimal, then it is recommended to install the sensor as close as possible to the centre of gravity in order to minimize the influence of parasitic bending moments on measurement;
   b. If the risk of buckling is present, then it is recommended to install the sensor far from the centre of gravity, but not on the cross-section principle axes; in this manner the sensor will be subject to high compression or tension in the case of buckling.
2. Since the strain in a column is constant along its length, the error due to the sensor's gauge length is minimal; consequently, it is recommended to maximize the sensor's gauge length for the following reasons:
   a. A longer gauge length better averages imperfections in inhomogeneous materials such as concrete
   b. A sensor with a long gauge length has a higher probability of detecting the damage that can occur at any location along the column
   c. The long gauge length better annuls the influences of parasitic bending moments for partially bi-encastered columns.
3. The temperature is to be monitored in columns exposed to significant environmental temperature variations in order to determine the thermal strain (provided that the thermal expansion coefficient of column material is known).
4. With the recommendations presented above, the following parameters are monitored:
   a. Load (normal force in the column), if the Young's modulus and thermal expansion coefficient are known, as well as cross-sectional properties of the column
   b. Young's modulus of column (if the load and cross-sectional properties of the column are known)

c. Average strain in the column
d. Deformation of column (relative vertical displacement between the column extremities)
e. Damage to column
f. Failure by compression and tension
g. Failure by buckling can be detected if the sensor is not installed in the centre of gravity nor on principal axes; if there is a risk of buckling, in order to avoid ambiguity in data analysis, it is recommended to equip the column with parallel topologies.

An example of a single topology applied to a column built as bending free in a zone with low temperature variations is presented in Figure 3.19. For a column built in environments with significant temperature variations, a temperature sensor is to be added in the middle of the sensor's gauge length.

A parallel topology is to be used for columns subject to bending. For reasons of simplification, the cases of uniaxial bending are presented here, but the same principles can be extended to biaxial bending using the considerations presented in Section 4.3.

Columns are structural elements that accept loads at their extremities. Thus, the bending moments can only be introduced at the column extremities and the moment distribution between the extremities is linear. Depending on the construction details, the bending can be introduced either at a single extremity of the column (column encastered only at one end) or on both extremities (bi-encastered column). The monitoring strategy in these two cases is different.

The column encastered at a single extremity has, at any given time, the maximum moment at this extremity and zero moment at the other extremity with a linear distribution of moments between the extremities (see Figure 5.8). The curvature distribution is linear (proportional to the bending moment distribution) only if the column's cross-sectional properties do not change along the column. In concrete columns, cracks may occur in the tensioned part of concrete (if any), in which case the cross-sectional properties change: the bending stiffness decreases. Consequently, the curvature distribution becomes nonlinear. Monitoring of the exact curvature distribution in the column requires the use of several parallel topologies, which is expensive.

The normal (axial) force in a column is expected to be negative (compression); thus, the part of the column with tension stresses is limited to the encastered extremity (see Figure 5.8), and the influence of cracks is expected to be limited too. Based on that consideration, the evaluation of a column's structural behaviour can be performed in a satisfactory manner using a single parallel topology. Considering the advantages presented above in points (2a)–(2c), it is recommended to install sensors that are as long as possible with the following reminders:

i. The middle point of the sensor is to be in the middle of the column height
ii. The maximum curvature in the column is equal (no cracks) or slightly bigger (cracks) than two times the curvature measured by sensors (see Figure 5.8).

The positions of sensors for a column encastered at a single extremity are shown in Figure 5.8.

Besides the parameters presented above in points (4a)–(4g), the following parameters are monitored using the strategy presented in Figure 5.8:

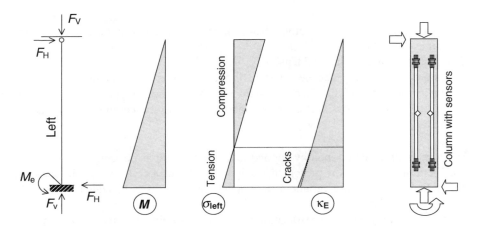

Figure 5.8 The strain distribution and sensor positions in a column encastered at a single extremity; the dashed line in the curvature diagram represents the distribution for noncracked material.

h. Estimation of maximal curvature and maximal average strain
i. Limited crack characterization
j. Estimation of bending moment and transverse force (limited accuracy in the case that cracks are present).

For a bi-encastered column the monitoring strategy is different, since the bending moments at both extremities are unknown (see Figure 5.9). Therefore, two parallel topologies are to be used. In order to determine the moments better and to avoid installation of sensors across the

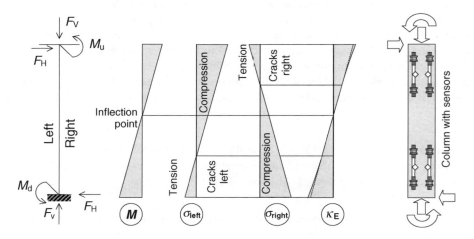

Figure 5.9 The strain distribution and sensors position in bi-encastered column; the dashed lines in the curvature diagram represent the distribution for noncracked material.

inflection point (whose position is, in general, unknown) it is recommended to use sensors with a shorter gauge length, typically equal to a quarter of the column length. An example of a bi-encastered column equipped with two parallel topologies is presented in Figure 5.9.

Monitoring of a column encastered at one end requires two times more sensors, and monitoring of a bi-encastered column requires four times more sensors than monitoring of an axially loaded column. The required number of sensors is even higher for biaxially bent columns. Consequently, the cost of monitoring is higher. In order to decrease the cost of monitoring it is recommended to evaluate the contribution of bending moments to the strain field in the column and, if this contribution is relatively small, the column might be considered as subject to a uniaxial load only.

5.3.3 Monitoring of Cores

A core is a stiff concrete structural member whose main purpose is to accept horizontal loads and torsion with respect to the vertical axis. Additionally, the core also accepts the vertical loads. The core commonly hosts elevators and staircases.

In a simplified manner, the core can be considered as a thin-walled vertical cantilever subject to all the possible loads: shear forces in horizontal directions, normal force, torsion and biaxial bending. Thus, the minimum recommended equipment necessary to monitor a single core cell consists of three crossed topologies (see Section 4.4.2) and a parallel topology for biaxial bending (see Section 4.3.2), amounting to nine sensors. In some simple cases this number can be lowered; in contrast, in some complex cases it must be increased.

Let a crossed topology be subject to a linear stress distribution, split in to its symmetrical and asymmetrical parts with respect to the vertical axis containing the crossing point, as shown in Figure 5.10. The symmetrical part of the stress distribution is equivalent to that generated by the normal force and the asymmetrical part is equivalent to that generated by the bending moment.

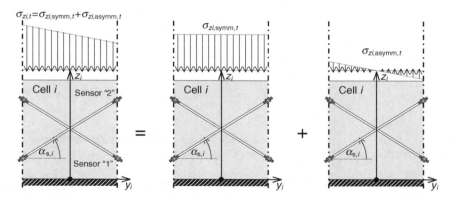

Figure 5.10 Part of wall subject to linear stress decomposed to symmetrical and asymmetrical components.

Supposing that stress in the horizontal direction (y_i axis) can be neglected and assuming linear behaviour of the material, then both sensors subject to symmetrical stress will measure the same value of strain, as follows:

$$\varepsilon_{i1,\mathrm{E_m},t} = \varepsilon_{i2,\mathrm{E_m},t} = \varepsilon_{zi,\mathrm{symm,E},t}(\sin^2 \alpha_{s,i} - v\cos^2 \alpha_{s,i}) \tag{5.1}$$

where v is the Poisson coefficient and $\varepsilon_{zi,\mathrm{symm,E},t} = \sigma_{zi,\mathrm{symm},t}/E_t$. Consequently, the average strain is determined thus:

$$\varepsilon_{zi,\mathrm{symm,E},t} = \frac{\varepsilon_{i1,\mathrm{E_m},t} + \varepsilon_{i2,\mathrm{E_m},t}}{2(\sin^2\alpha_{s,i} - v\cos^2 \alpha_{s,i})} \tag{5.2}$$

where the angle $\alpha_{s,i}$ must be selected in a manner that the following condition is fulfilled:

$$\sin^2\alpha_{s,i} - v\cos^2 \alpha_{s,i} \neq 0 \Leftrightarrow \tan \alpha_{s,i} \neq \sqrt{v} \tag{5.3}$$

The asymmetrical part of the stresses creates an asymmetrical strain distribution along the sensor. The integral of strain over the sensor's gauge length is zero, making the sensor insensitive to an asymmetrical stress distribution. Therefore, the value presented in Equation 5.2 is equivalent to the strain that could be measured with a simple topology containing a sensor installed through the crossing point. Thus, a spatial triangular topology can be considered as being equivalent to a parallel topology for biaxial bending.

Important factors for determination of sensor numbers and positions are the shape and dimensions of the core. The shape of the core can vary from a simple thin-walled rectangle to a very complex shape, with a number of openings, and the dimensions can range from a few metres to a few tens of metres, as shown in Figures 5.11 and 5.12.

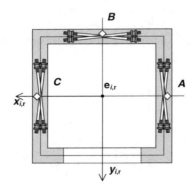

Figure 5.11   Plane (cross-sectional) view of a typical core, with simple rectangular shape and small dimensions; crossed topologies are presented only schematically.

A core with small dimensions and simple shape and whose behaviour can be considered as linear may be monitored with only three crossed topologies, as shown in Figure 5.11. In this case the same crossed topologies are used as an equivalent to the parallel topology presented in Figure 5.10 and Equation 5.2. The angles of the sensors are to be determined by taking into account the Poisson coefficient of the material (see Equations 5.1 and 5.2). If the core has an opening, then the sensors are to be installed in those walls without an opening, as shown in Figure 5.11.

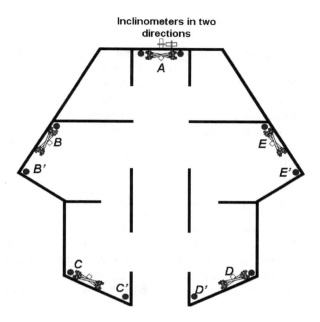

Figure 5.12   Plane (cross-sectional) view of a complex core, with important dimensions; crossed topologies are presented only schematically.

A core with bigger dimensions and a complex shape does not necessarily behave linearly. Openings and a horizontal dimension that is bigger than the distance between two storeys makes the strain field irregular in such a core. The equivalence between the crossed topology and simple topology may not be valid, and this is the reason for installing a separate parallel topology, as shown in Figure 5.12. Sensors in the parallel topology provide for a more accurate evaluation of the strain field in the core, but also for a certain redundancy, which is justified by the complexity of the core. Buildings with a complex and big core are usually very tall, representing landmarks in a city and hosting a significant number of people. These facts further justify the investment in the complete and redundant sensing system presented in Figure 5.12.

Monitoring the deformed shape of the core would be very expensive, since many cells are required to be equipped with sensors, and each cell contains a significant number of sensors. Monitoring can be reduced to the representative cells (i.e. to the most loaded cells). The bottom (ground floor) cell is certainly the most loaded and must be monitored. The selection of other cells along the core depends on the loads (e.g. a storey with an unusually high load) and the cross-sectional properties of the core (e.g. a dimensional change of the core on a particular storey). Besides deformation, it is recommended to monitor temperature and, in the case of very tall buildings, to monitor inclination of the core in both directions (see Figure 5.12).

With the strategies presented above, the following parameters are monitored:

a. Average strain in the core cell
b. Average shear strain in the core cell

   c. Global average curvature of the core cell
   d. Inclination of the core cell
   e. Approximate normal force in the core cell, if the Young's modulus, thermal expansion coefficient and cross-sectional properties are known
   f. Approximate shear force in the core cell, if the Young's modulus, thermal expansion coefficient and cross-sectional properties are known
   g. Approximate bending in the core cell, if the Young's modulus, thermal expansion coefficient and cross-sectional properties are known
   h. Damage to the core
   i. Failure by compression, tension and shear.

5.3.4 Monitoring of Frames, Slabs and Walls

A frame is a structural element containing a horizontal beam encastered to two columns at its extremities, while other columns may be present in between. The columns of the frame are monitored as shown in Section 5.3.2, while the horizontal beam, being subject to bending, is monitored using the necessary number of parallel topologies. Monitoring of beams subject to bending is presented in more detail in Section 5.4. A simple example is presented in Figure 5.13.

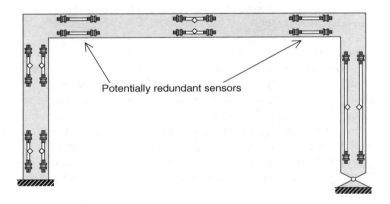

Figure 5.13   Schematic example of sensors network for frame monitoring.

   A slab is a structural element subject to biaxial bending and torsion, as well as normal and shear forces. Structural monitoring of a single slab cell requires at least six sensors (three parallel topologies), as shown in Figure 5.14. Consequently, a large number of sensors are required for whole slab monitoring, which is very expensive. In order to reduce the monitoring costs, it is recommended to monitor only the critical cells in the slab.

   A supporting wall is a structural element dominantly subject to the normal forces in its own plane, although it can be loaded with bending moments and torsion at its extremities. The bending moments and torsion (if any) can be monitored in the same manner as for the slabs. The normal forces can be monitored using a triangular topology. In order to minimize the number of sensors used, it is recommended to model the wall by the strut-and-tie method

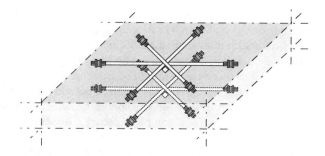

Figure 5.14   Schematic example of parallel topologies used for monitoring of slab cell.

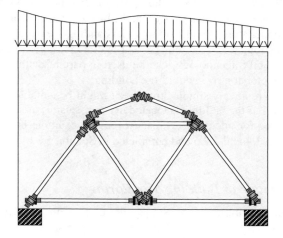

Figure 5.15   Schematic example of triangular topologies used for monitoring of walls whose internal forces are approximated by strut-and-tie method.

and to install the sensors in appropriate locations (i.e. where the struts and ties are situated). A simplified, schematic example is presented in Figure 5.15.

5.3.5 Monitoring of a Whole Building

An affordable monitoring system for global structural monitoring of a building contains a sensor network reduced to the most critical structural members; that is columns and core, and in particular the columns and core cells located at ground level. Horizontal beams and slabs are the structural elements that are, in general, not considered as critical – failure of a single beam or single slab is not likely to cause the failure of the whole building or loss of functionality of the whole building. However, it is important to correctly identify the most critical structural members and to even include in monitoring some slab cells and horizontal beams if necessary.

For very tall buildings with a complex structure the monitoring must include all the critical elements. Combined static and dynamic monitoring is recommended, regardless of the seismic environment. In order to determine the static structural behaviour it is necessary to equip the ground level with sensors (columns and core) and to repeat similar sensor constellations in several storeys throughout the building. The selection of the storeys to be equipped with sensors depends on the load distribution along the building height, changes in dimensional and material properties along the building height and the building height.

The deformation sensors are supposed to be read in both a static and dynamic manner and will provide for the dynamic properties of the building, such as natural frequencies and vibration modes. Thus, upper storeys equipped with sensors should be selected in such a manner to provide for these dynamic properties. The complete network of sensors should satisfy the requirements for both static and dynamic monitoring needs.

Beside the deformation sensors, it is recommended to install some other types of sensor in order to provide for complementary information and redundancy. Biaxial inclinometers at the ground level, as well as in some cells in the upper storeys, provide information concerning global rotation of the building. Biaxial or triaxial fibre-optic accelerometers installed in the basement and some upper storeys will help in understanding the dynamic loads acting on the building in the case of strong winds, earthquake or tremor. Finally, global movement monitoring systems (e.g. based on GPS technology) can be used in parallel with a fibre-optic system in order to determine the global movement of the building.

For buildings with a relatively simple structure it might be suitable to monitor only those columns at ground level. If the building is located in a non-seismic area, then static monitoring is suitable and provides for adequate results. An on-site example based on this strategy is presented in Sections 3.3.4 and 4.2.5 and completed in Section 5.3.6.

5.3.6 On-Site Example of Building Monitoring

The introduction and description of the Singapore building project (Glišić *et al.*, 2005) are presented in Sections 3.3.4 and 4.2.5. Some monitoring results are also presented below, with some additional clarifications of the project requirements and further data analysis being presented.

The monitoring strategy was developed based on the different criteria listed below:

1. Monitoring of critical members of the structure has been required. The critical members are structural elements in which malfunction or failure will generate a partial or even complete malfunction or failure of the structure.
2. Monitoring has to be performed at the local column level and at the global structural level. Knowledge concerning the behaviour of one or a few structural elements (columns) is not sufficient to make conclusions concerning the global structural behaviour; therefore, a representative number of elements had to be monitored.
3. The monitoring is to be performed over the whole lifespan of the structure, including the construction phase. The monitoring system selected for this type of monitoring must have an appropriate performance, notably high accuracy and long-term stability.
4. The monitoring system selected has to be designed for structural monitoring; it must not be influenced by local material defects in the concrete, such as cracks or air pockets.
5. The budget accorded to monitoring activities has been limited. Being a pilot project, which contains some uncertainties and which is subjected to development and changes, it was decided to limit the number of sensors installed in the building and to concentrate on the

results obtained from this limited number of sensors in order to evaluate the method and improve its performance.

6. For aesthetical reasons, it was not permitted for sensors and sensor cables to be visible or to egress directly from the columns.

The criteria presented have called for a particular monitoring strategy, including the selection of the monitoring system, the definition of the sensor type and position, the development of the installation procedures, the establishment of a measurement schedule and the development of algorithms for data analysis.

The conditions *sine qua non* for the selection of the type of sensor are imposed by criteria (3), (4) and (6). According to criterion (3), the sensor has to survive for long periods with high stability; hence, it has to be immune to corrosion, humidity, temperature variations and, electromagnetic field and interferences. According to criterion (4), the sensor selected has to have a long gauge. And according to criterion (6) it has to be embeddable in the concrete. Consequently, embeddable long-gauge fibre-optic sensors were selected for application.

A good compromise with respect to design criteria (1), (2) and (5) has been to equip the 10 most critical ground columns (between the first and second floors according to Singapore notation) with sensors (i.e. to use a scattered simple topology). The temperature in Singapore ranges between 23 and 33 °C approximately during the day or night, independent of the season. This fact, along with the limited budget for monitoring, led to the decision not to monitor the temperature. The ground columns were selected as being the most critical elements in the building, and the number of sensors was adapted to the available budget. The positions of the columns monitored in the ground floor plan are given in Figure 4.14. The positions of the sensors within the column and the photographs taken during the installation are presented in Figure 3.19. The views to the monitored building during and after construction are shown in Figure 5.16.

Figure 5.16 The 19-storey-tall residential building in Singapore monitored during construction (left) and after completion (right) (courtesy of SMARTEC and Sofotec).

The monitoring schedule included the most important phases of the structure's life: construction of 19 storeys, periodic measurements, periodic 48 h continuous monitoring sessions and sessions after unusual events, such as tremor. The results collected during the first 5 years with different monitoring sessions are presented in Figure 3.20.

Each column was first analysed at the local level in order to compare the measured values with the design values. An example of this comparison is presented in Section 3.3.4 and in

Figure 3.21. Unusual behaviour in the form of drift of measured strain from design values is noticed in several columns in the period 17 September 2002–4 April 2003. The drift was estimated to be a consequence of differential settlement of the foundations (see Section 4.2.5). Simplified algorithms allowed estimation of the differential settlements to be ranged between 0.25 and 0.33 mm, with the exception of columns neighbouring C1 and C3, whose drift was persistent and ranged between 0.5 and 1 mm. The differential settlement detected is rather low and not dangerous; however, the monitoring system and algorithms applied demonstrated adequate performance in detecting it early.

The aims of the 48 h monitoring campaigns performed regularly since July 2004 were (1) to learn about the building's behaviour caused by daily temperature changes and inhabitant fluctuations and (2) to record the health state of the building as a reference for comparison with future monitoring results.

It was decided to perform one measurement over all the sensors every hour in order to monitor the structural behaviour of the building quasi-continuously. Embedded temperature sensors were not present; therefore, only ambient temperature was monitored, along with relative humidity and a weather description, in order to observe possible correlations. Averaged values of temperature with their minimum and maximum variation for each campaign are presented in Figure 5.17. Relative humidity was constantly at 75 %, with variations of ±10 % for each campaign.

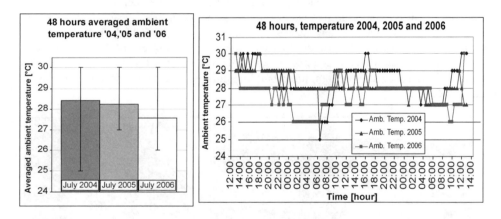

Figure 5.17   Ambient temperature during the 48 h campaigns (data provided by Sofotec, HDB and SMARTEC).

Since the 48 h period is relatively short, that part of the strain generated by creep and shrinkage can be neglected. Thus, the two main factors influencing the average strain variations are temperature variations in the columns due to weather and day–night heating and cooling, and vertical load changes due to movement of inhabitants that are at work during the day (empty building, lower load) and at home during the night (full building, higher load).

The values of total measured strain are relatively stable during the observed periods and show a dependence rather on ambient temperature variations, since the load changes were small (not all the apartment are inhabited). This dependence is different for different columns: for some columns it is small and for some other columns it is a little bit higher.

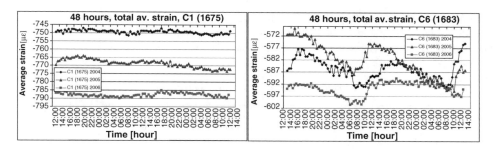

Figure 5.18 Total average strain in columns C1 and C6 recorded during the 48 h continuous monitoring sessions performed in July 2004, July 2005 and July 2006 (data provided by Sofotec, HDB and SMARTEC).

This different reactions of columns can be explained by different column temperatures with respect to the ambient temperature and interaction with other columns and the storey slab. Temperature variations in columns are smaller or bigger than ambient variation due to thermal inertia of the concrete and direct exposure to the sun. Examples of strain variations in a column not directly exposed to the sun and in a column directly exposed to the sun are given in Figure 5.18.

In the evening hours (18:00–22:00) the temperature was relatively constant, but inhabitants are expected to come home and load the building. Since no significant compression in the columns was recorded during this period, it is reasonable to conclude that the in-and-out movement of inhabitants does not influence the behaviour of the building significantly.

The values of total average strain recorded from year to year are expected to increase in absolute value due to creep and shrinkage. As mentioned above, columns C1 and C3 experienced a drift from their design values and the 48 h campaign confirmed this. The summary of the column averaged strain evolutions from 2004 to 2006 with their maximum and minimum variations is presented in Figure 5.19.

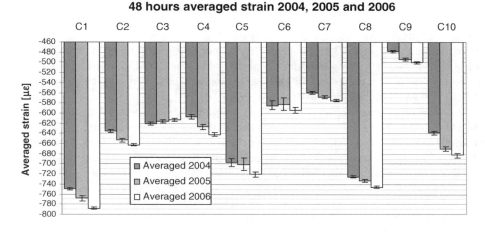

Figure 5.19 Summary of averaged 48 h campaign strain evolution in the period 2004–2006 (data provided by Sofotec, HDB and SMARTEC).

In March 2005, an earthquake in neighbouring Indonesia created a tremor in Singapore. In order to evaluate potential degradation in structural performance, a single session over all the sensors was performed just after the tremor. The results of this session are presented in Figure 5.20.

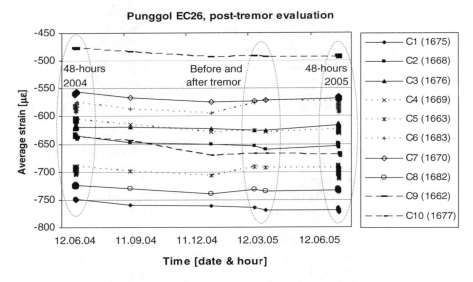

Figure 5.20   Average strain measurements recorded before and after a tremor (data provided by Sofotec, HDB and SMARTEC).

The change in strain before and after the tremor varied in different columns, from −7 to +5 $\mu\varepsilon$. This variation is considered to be regular, and generated by temperature and live load variations rather than by tremor, since it was in the range of the 48 h variation registered in 2004. Moreover, the range of 48 h variation of strain registered in 2005, which is similar to that registered in 2004, confirmed that no degradation of performance occurred due to the tremor.

A pioneer project for the monitoring of residential buildings in Singapore is presented in this section and Sections 3.3.4 and 4.2.5. The monitoring strategy and the results collected during 5 years on a 19-storey building are presented and analysed. The registered parameter was average strain in columns and this allowed the monitoring of structural behaviour at a local column level and a global structural (storey) level.

The use of fibre-optics sensors on such a large scale for monitoring of high-rise buildings is the first in Singapore and sets directions that will help designers to better understand the behaviour of tall buildings during their life cycle from construction to service conditions.

Such pioneering efforts have already yielded results from the insights gained from enlarged knowledge concerning the real behaviour of columns during construction. Differential settlement that is of low magnitude and not dangerous (at the present time) to the building's performance was detected. The 48 h sessions allowed better assessment of the building's performance in the long term and made possible a post-tremor analysis.

The monitoring strategy employed and the fibre-optic monitoring system selected have successfully responded to the design criteria. The monitoring strategy has shown high performance in spite of the limitations imposed by the design criteria (limited number of columns equipped, lack of temperature measurement, lack of accurate shrinkage and creep coefficients, uncertainty concerning the real load during measurement campaigns, etc.).

5.4 Monitoring of Bridges

5.4.1 Introduction

Bridges are probably the most attractive civil structures. In construction they range from simple beams with spans of few tens of metres to suspended bridges with a span easily exceeding 1 km. They can be built from any of commonly used construction materials: wood, masonry, steel, concrete and, recently, composites.

The importance of bridges as a means of communication has long been appreciated. That is why particular care is given to the maintenance of bridges with the aim to keep them functional as long as possible. Nowadays, there are some very old bridges in service, with the list of all their maintenance works engraved in stone, written in different languages and performed by different successive authorities, confirming this human attitude through different times and cultures.

SHM certainly contributes greatly to maintaining bridges in service. Fibre-optic sensors with a long-gauge basis, offering the possibility of global structural monitoring, made possible the creation of basic monitoring strategies for bridge monitoring based on the bridge structural system. These monitoring strategies will now be presented in a general manner. They are not to be considered as immutable solutions, but rather as the basis for the creation of a more detailed monitoring strategy, depending on the project requirements.

The monitoring strategy for masonry bridges was presented in Section 4.5.2. For other construction materials, the general monitoring strategies are presented in the following sections for a simple beam, a continuous girder, a balanced cantilever, an arch bridge, a cable-stayed bridge and a suspended bridge. When possible, the monitoring strategy is illustrated with an on-site example.

5.4.2 Monitoring of a Simple Beam

A simple beam is a linear element supported at its extremities by two abutments and subject to bending, uniaxial or biaxial, as well as to shear forces. In the case of prestressed concrete, the beam is subject to normal forces; and if the beam is skewed or subject to asymmetric traffic load (e.g. traffic jam in one direction), then torsion can also occur.

A general solution requires the use of parallel topologies for normal efforts and uniaxial or biaxial bending combined with crossed topologies for shear forces and torsion. Since a simple beam is an isostatic structure, inclinometers will help in monitoring of differential settlements of abutments and the evaluation of torsion. An example of a general monitoring strategy for a skewed simple beam is presented in Figures 5.21 and 5.22 Figures 5.21 (plan view) and 5.22 (cross-sectional views).

A particular problem in the detection of a structural malfunction of a simple beam is its isostatic structural system. When damage occurs in a non-instrumented zone (between two cells), the sensors cannot detect the damage directly, nor indirectly – the static system due to damage does not change and the strain distribution in the beam does not change. Assuming that

Figure 5.21   Plan view of example of monitoring strategy for skewed simple beam (courtesy of SMARTEC).

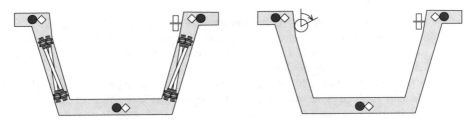

Figure 5.22   Position of sensors in cross-section close to left abutment and in the middle of the span (courtesy of SMARTEC).

damage is most likely to occur in the most stressed cells, it is rational to equip these cells with sensors. The stress in the cell depends on internal forces and moments, but also on geometrical and mechanical properties of the cell, thus all of them are to be considered when developing a monitoring strategy.

In addition, attention is to be paid to positioning of the sensors. Improper positioning of a sensor can lead to wrong interpretation of data and, consequently, the global performance of a monitoring system can be compromised. Taking into account the above guidelines, the sensor network for a simple beam is developed as follows.

The sensors making up the parallel topology must be installed in the same cross-section; that is, the centres of the sensors are to be installed in the plane perpendicular to the elastic line of the beam. Only a sensor installed in this manner can provide for accurate measurement of curvatures and central (axial) strain.

The minimum number of cells to be equipped with a parallel topology depends on the beam's complexity. If the beam is prestressed, then it is necessary to install at least one parallel topology in the middle of the span (the most loaded cell) and one at each extremity. The topologies on the extremities are necessary in order to correctly determine boundary conditions that are influenced by the intensity and eccentricity of the prestressing force. A similar consideration is valid for a skewed beam, whether it is prestressed or not – due to geometry, bending moments and curvatures are generated near to extremities. Only in the case of a straight simple beam without prestressing can sensors at the extremities be skipped, considering that the curvatures at the extremities can be neglected. Besides the locations of the parallel topologies presented above, it is necessary to install more topologies practically in any other cell that is subject to high loads (e.g. reduction of cross-section, change in reinforcing or prestressing).

However, in order to increase the probability of detecting an unusual behaviour in the simple beam, to perform more accurate data analysis and, finally, to provide for a certain redundancy, it is recommended to install more parallel topologies than the minimum recommended number. For example, for a simple beam with constant cross-section, additional topologies can be installed in the quarter spans, as shown in Figure 5.21. Parallel topologies provide for the following data:

a. Determination of strain field in the beam
b. Distribution of average central strain
c. Distribution of average horizontal and vertical curvatures
d. Crack characterization of concrete, if the sensors were embedded
e. Determination of horizontal and vertical deformed shape
f. Evaluation of normal (axial) force, if cross-sectional properties, Young's modulus and thermal expansion coefficient are known
g. Evaluation of bending moments, if cross-sectional properties, Young's modulus and thermal expansion coefficient are known
h. Determination of prestress losses, indirectly through monitoring of concrete
i. Damage detection, assuming that it occurs on sensors and that the sensors are installed at the most critical locations.

The dominant shear strain generated by vertical shear forces is expected to occur near the abutments. That is the reason for equipping the cells at the extremities of the beam with crossed topologies, as presented in Figures 5.21 and 5.22. In common practice, the shear strain generated by a horizontal shear force can be neglected – the main purpose of beam bridges is to accept vertical loads. Thus, only one crossed topology is sufficient to determine the shear strain near to an abutment.

Torsion is rarely an issue in a simple beam. It can occur due to shape (skewed beam), but also due to eccentricity of the load (e.g. traffic jam causes a queue in one lane). The biggest torsion occurs near to the extremities; therefore, if torsion is an issue, crossed topologies are to be installed at the extremities of the beam (see Figures 5.21 and 5.22). Since the shear strain can also be generated by shear forces, it is necessary to install at least two crossed topologies in order to separate the influences (see Section 4.4). The crossed topologies provide for the following data:

a. Determination of shear strain in the cell equipped with sensors
b. Determination of shear strain generated by shear force
c. Determination of shear strain generated by torsion, if the cross-sectional properties of the cell are known
d. Evaluation of shear force, if the cross-sectional properties and shear modulus are known
e. Evaluation of torsion moment, if the cross-sectional properties and shear modulus are known
f. Damage detection at the extremity of the beam.

To determine the thermal strain component it is necessary to monitor temperature. Since the temperature commonly does not vary very much along the beam, it may not be necessary to install temperature sensors in each cell (see Figure 5.21). A reduction in the number of temperature sensors lowers the cost of the monitoring system. The cells equipped with a temperature sensor must be equipped with at least three sensors in order to provide for thermal gradients (between the sunny and shadow sides of the girder, east and west, top and bottom, etc.).

Bending provokes rotations of cells that do not necessarily have to be monitored directly – good data are provided with parallel topologies. Nevertheless, the movements of the bridge, the differential settlements of abutments and damage in the form of a big crack between cells cannot be detected by deformation sensors alone. That is why it is recommended to install at least one inclinometer that monitors rotation in the vertical–longitudinal plane. The cells that rotate the most due to bending are those at the extremities. The cell that rotates the least is the one in the center of the span. An inclinometer installed at an extremity of a beam is sensitive to both rotation generated by bending and rotation generated by differential settlements of abutments and/or damage. On the other hand, an inclinometer installed in the middle of a span is mainly sensitive to differential settlements of abutments and/or damage. The selected position depends on the project specifications. Installation of both inclinometers certainly provides for redundancy, more accurate results and more complete results. Inclinometers provide for the following data:

a. Rotation in cells equipped with inclinometers
b. Distribution of vertical displacements (combined with parallel topologies)
c. Detection of differential settlements of abutments
d. Detection of damage in the form of a big crack that occurs between cells equipped with deformation sensors.

For simple beams with torsion, the biggest rotation generated by torsion occurs in the middle section; that is the reason to select this section for monitoring and to equip it with one inclinometer that monitors the rotation in the vertical perpendicular plane. This inclinometer monitors:

a. Torsion generated rotation in cell equipped with inclinometers
b. Damage detection due to torsion.

A simple beam bridge can be supported at two extremities by columns that can be founded on piles. A general monitoring strategy for these structural elements is presented in the previous sections, and similar principles can be applied in the case of bridge columns and foundations.
    An on-site example of monitoring of a simple beam is presented next.

5.4.3 On-Site Example of Monitoring of a Simple Beam

An introduction to the monitoring project of interstate highway bridge I-10, close to Las Cruces, New Mexico (Idriss and Liang, 2006), along with some results, is given in Section 4.4.3. The presentation of the project is further developed and completed in this section.
    The interstate highway bridge I-10 consists of twin bridges (two parallel bridges) with five spans each (Hughs, 2004). A span consists of six simple skewed beams, pre-fabricated and prestressed in an open-air plant. The beam cross-section is 1.37 m tall and has a 'U' shape with wings, as shown in Figure 5.21. The beams are built of high-strength, high-performance prestressed concrete with a minimum 28-day strength of 68.9 MPa.
    The prestressing strands were first put in tension, then the formwork was positioned and the rebar cages were assembled. Finally, concrete was poured and the beams were steam cured at a temperature of approximately 60–90 °C for at least 2 days, in order to accelerate the maturation of concrete. When the tests proved that the concrete had achieved its designed

strength, the strands were cut and the beam prestressed; formwork was then removed and the beams transported to the storage area.

After all the beams were fabricated, they were transported from the storage area to the bridge site, placed on the columns and the deck was poured over the beams. The scheme of the bridge is presented in Figure 5.23.

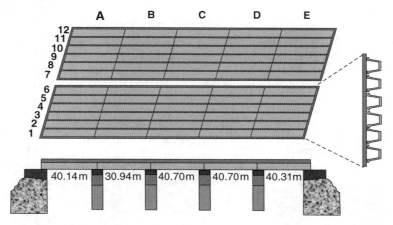

Figure 5.23   The schema of the I-10 bridge close to Las Cruces, New Mexico (courtesy of SMARTEC data provided by New Mexico State University).

The numerical models for high-performance concrete were not established in codes and the models for ordinary concrete were used. The main aims of monitoring were to increase knowledge of the structural behaviour and to verify performance and indirectly to ensure safety. The monitoring focuses on determination of long-term prestress losses (Liang, 2004), and primary monitoring parameters were average strain, average shear strain, average curvature and deformed shape. For this purpose it was decided to equip beams A1–A6 with deformation sensors, as presented in Figure 5.24.

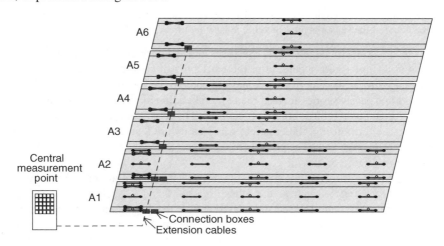

Figure 5.24   Position of sensors in monitored beams of I-10 bridge with connection schema (courtesy of SMARTEC data provided by New Mexico State University).

Beams A1 and A2 were fully equipped with sensors following the strategy presented in Section 5.4.2, with few restrictions imposed by economical issues: the torsion and differential settlements were not of primary interest and that is why the inclinometers were not used; only the cell near to the abutment was equipped with crossed sensors.

Having beams A1 and A2 well equipped with sensors, in order to reduce the cost of monitoring it was decided to equip other beams with less sensors. The measurements of the sensors installed in beams A3–A6 were compared with those registered with the corresponding sensors installed in beams A1 and A2. The 2 m long sensors were used. The sensors were attached to the rebars using plastic ties. The ties only straighten out the sensor, but do not establish a rigid connection between the sensor and the rebar. In this manner, the sensors were embedded in the fresh concrete, and measured the strain in the concrete. The sensors installed on the rebars are shown in Figures 5.25.

Figure 5.25 Sensor installed on rebar cage (black arrows) and cables leading to connection box (white arrows); the detail of the sensor is presented in the inset (courtesy of SMARTEC).

All the sensor connecting cables (passive zones) were also attached to the rebars and led out of the beam at a single point where, after the hardening of concrete, the connection box was installed (see Figure 5.24). The passive zones attached to the rebars are shown in Figure 5.25.

The sensors and passive zones are designed for embedding in concrete; thus, the concrete was poured and vibrated using the common procedures. The rebars provided additional protection for the sensors. After pouring, the concrete was steam cured, strands were cut and the beam transported to the storage area. The pouring of the concrete and the finished beam in the storage area are shown in Figure 5.26.

Monitoring started immediately after the pouring. The early and very early age strain of the beam was registered, as well as the other important stages, such as prestressing, transportation to storage area and long-term behaviour. The results obtained during the first days after the pouring for the crossed topology are presented in Section 4.4.3. An example of the results obtained with the parallel topology installed in the middle of the beam is presented in Figure 5.27.

Figure 5.26 (a) Pouring of concrete over the sensor (sensor indicated with white arrows) and (b) beam in storage area with installed connection box (white arrow) connected to reading unit (not visible, behind protection tent) using connecting cable (black arrow) (courtesy of SMARTEC).

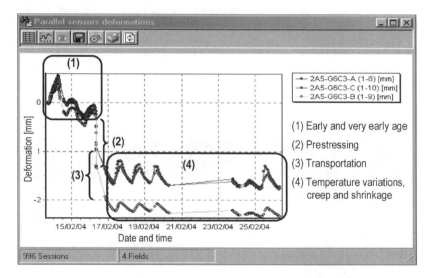

Figure 5.27 Typical results obtained on I-10 bridge by sensors forming parallel topology in the middle of the span (courtesy of SMARTEC data provided by New Mexico State University).

The temperature variations registered in the middle of the beam are presented in Figure 5.28.

The measurements performed during the first 12 days are presented in Figures 5.27 and 5.28. Four different stages are registered: (1) early and very early age, (2) prestressing, (3) transportation to storage area and (4) behaviour in the storage area.

The dominant strain evolution during the early and very early age arises from the temperature generated by the hydration process and steam curing. During steam curing the beam was covered with insulating mats. The beam was in contact with the formwork, which was in contact with cold soil that was taking away the heat. This is confirmed by the lower temperature measured

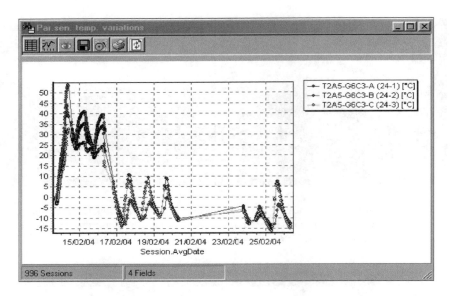

Figure 5.28 Typical temperature evolution obtained on the I-10 bridge by temperature sensors in the middle of the span (courtesy of SMARTEC data provided by New Mexico State University).

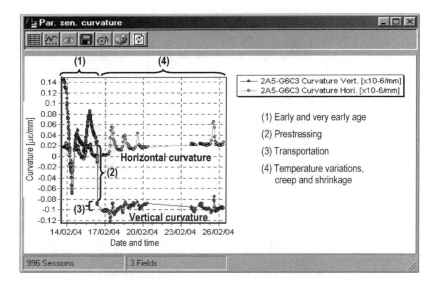

Figure 5.29 Horizontal and vertical curvature generated by the (prestressing) forces and creep (courtesy of SMARTEC data provided by New Mexico State University).

with the bottom temperature sensor and the lower strain evolution in the bottom part of the beam. Hence, after this stage was completed, before prestressing, the beam was already subject to bending generated by thermal processes.

Prestressing introduced a high eccentric normal (axial) force that generated shortening and bending of the beam in the vertical plane. The sensor installed in the lower part of the cross-section deformed more than the sensors installed in the upper part. The horizontal and vertical curvatures due to forces and creep in the middle cell of the beam were calculated using the equations presented in Section 4.3 and presented in Figure 5.29.

After the prestressing, the vertical curvature is clearly observed while the horizontal curvature is equal to zero, indicating symmetrical redistribution of the prestressing force in the beam's cross-section with respect to the vertical principal axis. Transportation only slightly increased the curvature in the vertical plane.

The deformation of the girder in the storage area was mainly influenced by temperature variations that generated internal stresses, notably due to the different exposures to sun of two sides of the beam and thermal inertia of the material. A small horizontal curvature was also generated and stabilized after 1 week.

Double integration algorithm was applied in order to determine the camber. The value of 82.5 mm was obtained while direct measurements have shown 85 mm, confirming good agreement between two measurements methods.

Two months after the fabrication, the beams were transported to the site and laid over the columns. The sensors from all the beams were connected via extension cables to the central measurement point. The cables were guided through an underground conduit in order to protect them in the long term. The central measurement point was built on-site, approximately 15 m far from the bridge. It has been provided with electrical power and air-conditioning system.

All the beams were connected to the central measurement point in parallel with installation of reinforcement cages on the deck. In this manner, the monitoring system was in a fully operating condition before the pouring of the deck and ready to register deformation due to the dead load of the deck. From this moment on, the measurements were taken regularly, every hour.

Figure 5.30   View of central measurement point (black arrow) and beams near abutment; connection boxes are indicated with a white arrow and extension cables with a grey arrow (courtesy of SMARTEC).

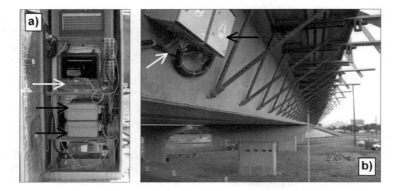

Figure 5.31 (a) View of equipment in central measurement point, reading unit (SOFO type, indicated with white arrow) and channel switches (black arrows); (b) beams installed on columns with connection boxes (black arrow) and extension cable (white arrow) (courtesy of SMARTEC).

The view of the central measurement point is given in Figure 5.30 and the contents of the central measurement point and the beams installed on the columns are shown in Figure 5.31.

After pouring of the deck, the beams were deformed and the camber changed. The deformed shape of the girder before pouring the deck is shown in Figure 5.32 and the evolution of the camber before and after pouring the deck is shown in Figure 5.33.

Monitoring was performed over the long term and, in particular, the prestressing losses were determined from concrete deformations. Monitoring of concrete deformation provided for total prestressing losses, including those generated by relaxation. The losses generated by relaxation could not be fully measured if the sensors were installed directly on the strands, since the relaxation occurs without dimensional changes. The monitored prestressing losses are compared with different numerical models found in codes and the literature in Figure 5.34.

Since numerical models were not established for high-performance concrete, the results of modelling presented in Figure 5.34 were obtained using numerical models for ordinary

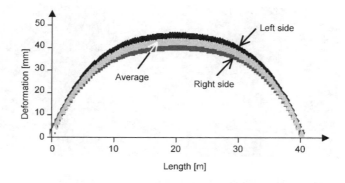

Figure 5.32 Deformed shape of girder before pouring of deck (Idriss and Liang, 2006).

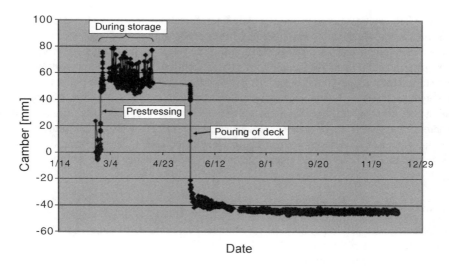

Figure 5.33 Typical evolution of camber before and after the pouring of the deck (Idriss and Liang, 2006).

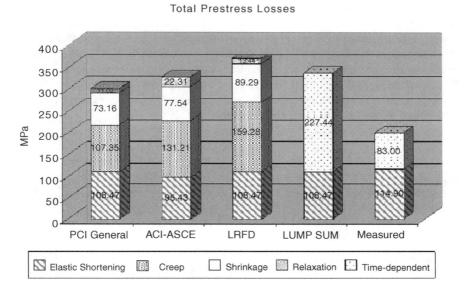

Figure 5.34 Comparison between prestressing losses calculated using different numerical models and those measured with long-gauge fibre-optic sensors (Idriss and Liang, 2006).

concrete. All the models found in codes and the literature predicted higher losses than measured. Lower measured losses can be explained by lower creep in high-performance concrete.

In the presented project, the fibre-optic monitoring system with long-gauge sensors provided rich information on the structural behaviour of beams built of high-performance concrete. Monitoring started immediately after the pouring of concrete and all important phases of the beams' lives were registered: early and very early age deformation, prestressing, transportation, installation on site, pouring of the deck and in-service performance. Besides the primary monitored parameters (such as average strain, average shear strain and average curvature), deformed shape, the influence of torsion, vertical camber evolution and, most importantly, prestressing losses were calculated. Comparison with the models confirmed the safe behaviour of the beams.

5.4.4 Monitoring of a Continuous Girder

A continuous girder is a hyperstatic linear element that consists of single beam supported on abutments at its extremities and by several columns in between. The part of the girder between two supports is called the span. A continuous girder is, in general, subject to the same influences as a simple beam: bending, uniaxial or biaxial, shear forces, normal force (prestressed concrete) and torsion (skewed spans, transversal rotation of supporting columns, asymmetric load). An example of a continuous girder bridge is given in Figure 5.35.

A general solution for each span of a continuous girder is similar to those presented for a simple beam in Section 5.4.2. Thus, a set of parallel and crossed topologies is to be combined with inclinometers and thermocouples. A typical sensor network for a continuous girder is presented in Figure 5.35.

The continuous girder is supposed to have longer spans than a simple beam and is subject to negative bending moments in the parts above the supporting columns. It is often shaped according to the bending moment's distribution: the cross-section in the middle of the span has smaller height than that located above the supporting column. The cross-section, therefore, is not constant along the girder and its elastic line is not straight. In order to make possible the application of the algorithms presented in Chapter 4 it is necessary that sensors forming parallel topologies are also parallel to the elastic line. This condition has important implications for the positioning of the sensors in the cross-section.

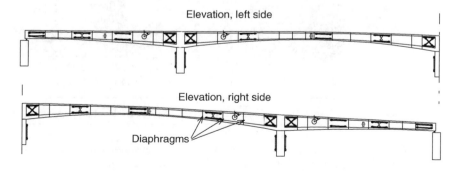

Figure 5.35 View of continuous bridge with proposed sensor network (courtesy of SMARTEC).

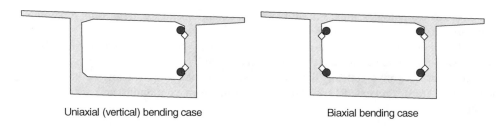

Uniaxial (vertical) bending case   Biaxial bending case

Figure 5.36   Cross-sectional position of sensors in girders with variable cross-section (courtesy of SMARTEC).

The cross-section of continuous girders is commonly built in the form of a box (box girder). Thus, the parallel topology sensors, in order to be also parallel to the elastic line of the girder, must be embedded or installed on the lateral walls. For cases where only vertical bending is of interest, the ideal solution would be to install the sensors over the principal vertical axis, but this cannot be respected – also, for this case the sensors are to be installed in the lateral wall. Besides the correct interpretation of the measurement of the parallel topology, the installation of sensors in the lateral walls provides for better protection of the sensors, since they are not directly exposed to potential contact with the maintenance team inspecting or repairing the interior of the bridge. The proposed position of the sensors of the parallel topology in the cross-section is for uniaxial and biaxial bending, as presented in Figure 5.36.

Another issue may be in attendance when sensors are to be installed on the surface of a structure: the box girders are stiffened against torsion by diaphragms that present physical obstacles for the sensor. The length of the sensor should not exceed the distance between diaphragms, otherwise the diaphragms must be drilled, which requires expensive tools and is time consuming.

The continuous girder is a hyperstatic structure. Consequently, a malfunction of the structure caused by damage (cracks) causes a change of the deformed shape of the girder, which is detected by deformation sensors. The frequent problem of continuous girders is differential settlements of supports – columns and abutments. This phenomenon also causes a change in the deformed shape and is detected by deformation sensors. Therefore, the inclinometers that measure rotation in longitudinal–vertical plane are not necessary, but they are still recommended because they provide for better accuracy and redundancy. These inclinometers are to be installed either close to the inflection point (biggest rotation see Figure 5.35) or in the middle spans (smallest rotation) as, discussed for the case of a simple beam.

Alternative methods for detecting the differential settlements of the supports are the monitoring of columns (indirect) and the use of an optical-fibre hydrostatic-levelling system (direct). Depending on the importance of the structure and the available budget, one or both of these alternatives can be used along with deformation sensors to provide better accuracy and redundancy.

In some cases, where the bridge consists of short spans supported on a large number of columns, it is rational to monitor only differential settlements using one or both alternatives. The spans, being short, may not necessitate deformation monitoring.

In the horizontal plane, the elastic line can be straight or curved. The spans of curved girders can be equipped with a similar sensor network as the straight girders. In addition curved girders are subject to very strong torsion, and so sensors related to torsion monitoring are to be used;

that is, crossed topologies at the extremities and an inclinometer for monitoring rotation in the transverse plane installed in the middle of the span.

The foundations of the columns and the foundations of the bridge can, in general, be monitored using the strategies presented in Sections 4.2 and 4.3. The monitoring parameters for continuous bridges are practically the same as in the case of a simple beam (see Section 4.5.2); hence, they are not repeated here. An on-site example of continuous bridge monitoring is presented next.

5.4.5 On-Site Example of Monitoring of a Continuous Girder

The basic characteristics of the Versoix Bridge in Switzerland are presented in Section 4.3.4 (Vurpillot, 1999; Inaudi *et al.*, 1999). The bridge combines existing and new concrete and monitoring was employed in order to evaluate its performance: to see whether it behaves as a single monolithic structure or rather as two separate structures, an old one and a new one. Views of the Versoix Bridge under construction are presented in Figure 5.37.

To obtain a good representation of the bridge deformations it was necessary to find a means to monitor the horizontal and vertical deformed shape of the bridge during the different construction phases and in the long term. In order to guarantee a sufficient redundancy and to follow the concrete pouring stages of 14 m, one section of sensors was placed every 7 m. Because of budget limitations, only two of the six spans were instrumented and only parallel topologies were used. A preliminary study by finite element simulation showed that the deformations of the two spans could be approximated by two fifth-degree polynomial functions. At least four curvature measurements for each span are therefore, necessary, to obtain the bridge spatial displacements. In each section, five sensors were embedded in the new concrete and one sensor was installed on the surface of the old concrete (see Figure 4.32).

To obtain a good representation of the average curvature, sensors with a 4 m active length were chosen. Two additional sensors were installed to give information about the differential shrinkage between the old and the new concrete. These sensors were rigidly connected to the existing concrete and their measurements could be compared with parallel sensors placed in

Figure 5.37   Versoix bridge under construction (Vurpillot, 1999).

the newly added concrete. This installation scheme was repeated 12 times: five times in the first span and seven times in the second. The horizontal and vertical curvature is calculated separately for each of the 12 sections using all eight sensors. The spatial displacements are then calculated by integrating the mean curvature of the 12 sections. The cross-sectional positions of the sensors are given in Figure 4.32 and a plan view of the sensor network is given in Figure 5.38.

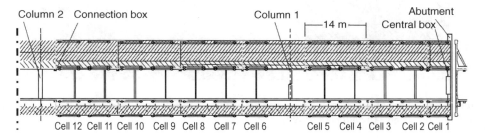

Figure 5.38   Plan view of the sensor network on Versoix Bridge (Vurpillot, 1999).

To facilitate the implantation of automatic and remote monitoring, the whole sensor network had to be measured from one single and easily accessible location: the abutment. The main sensor network is composed of 96 fibre-optic deformation sensors with a 4 m active length and 2–10 m passive length, 14 optical cables with 10–100 m length, 14 local connections boxes and one central box. The central box also allows the installation of a reading unit, optical switches, portable PC and modem to measure the bridge remotely. Views of the connection box and central box are presented in Figure 5.39.

The sensor installation followed the bridge widening schedule. The installation was very rapid: 2h was sufficient to place four sensors in each concrete pouring stage (for the interior overhang widening). Sensors were attached to rebars just after their completion and the building

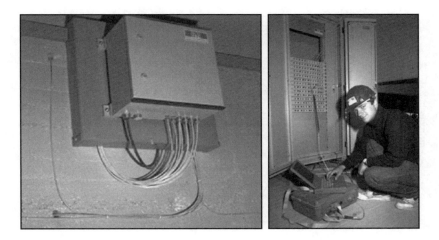

Figure 5.39   View of connection box (left) and central connection box (right) (courtesy of SMARTEC).

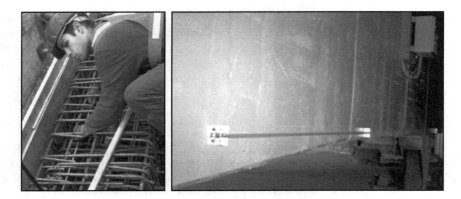

Figure 5.40 View of installation of sensors by embedding (left) and surface mounted sensor with connection box (right) (Inaudi *et al.*, 1999a).

yard schedule was not delayed. Sensors were only held with plastic rings (not fixed) to the rebars. Connection boxes were placed at the same time as the sensors in order to protect the optical connectors, to check each one during the installation and to measure the bridge during construction. Finally, before the catwalks allowing access to the local boxes were removed, all boxes were linked to the central box. The exterior sensors were attached to the surface of the old interior web using L-shaped metallic adapters. Installation of sensors to the rebars and the completed surface installation are shown in Figure 5.40.

The typical deformation evolution during the first 8 months is presented in Figure 5.41. The deformation sensors embedded in concrete belonging to the same pouring stage indicated approximately the same behaviour. Different strain components are present: elastic strain, creep, shrinkage and thermal variations.

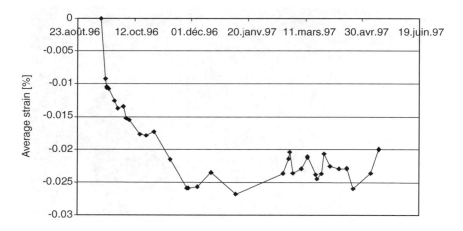

Figure 5.41 Typical evolution of detormation during the first 8 months (Vurpillot, 1999).

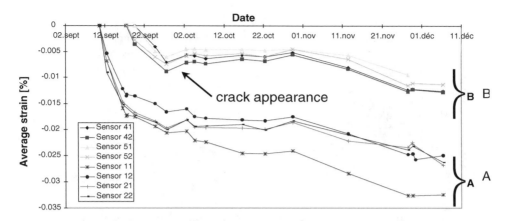

Figure 5.42 Comparison between the average strain evolutions for pouring stages A and B and detection of crack apparition (Inaudi *et al.*, 1999a).

Measurements performed on two successive concrete pouring stages, A and B, are shown in Figure 5.42. The sensors in the same stage registered similar results, but the two groups showed different behaviour. Stage A had shrinkage of 0.02 % approximately, while the stage B had an apparent shrinkage of only 0.005 % approximately. This difference is explained by the use of different concrete mixtures between the two stages and the different climatic conditions. Stage B was subject to cracking whereas stage A was not. The shrinkage in stage B is practically entirely compensated by the crack openings, and this explains the lower values of shrinkage. The monitoring system provided for an easy and early detection of crack onset.

An example of the monitoring of the interaction between the old and new concrete is presented in Figure 5.43. The measurements of shrinkage for two parallel sensors, the first installed in the new concrete and the second installed in the old concrete, are compared. The two curves

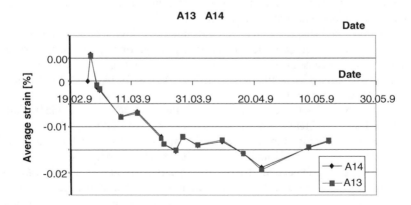

Figure 5.43 Comparison between the average strain evolutions of sensors installed in old and new concrete (Inaudi *et al.*, 1999a).

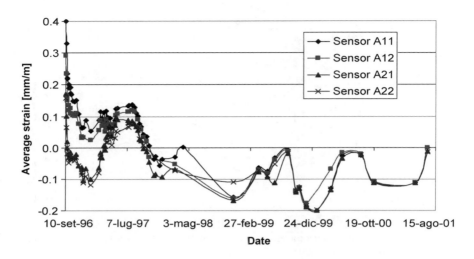

Figure 5.44 Average strain evolution in a Versoix bridge cross-section during the 5-year period (courtesy of SMARTEC).

match very well, indicating an overall differential shrinkage close to zero, and good cohesion between the two concretes.

A typical long-term evolution of the sensors installed in the same cross-section is presented in Figure 5.44. In order to compensate the very early age deformation, which occurs before hardening of concrete and without significant stress generation, all the measurements were set to the same value at a later stage (after 4.5 years approximately). Big differences were noticed during the first year, whereas later on the behaviour of the sensors became more similar. This behaviour can be explained by equilibrium of rheologic deformations in the new and old concretes.

The horizontal and vertical deformed shapes of the first two spans of the bridge were calculated using the double-integration algorithm. Horizontal deformation was caused by asymmetrical shrinkage of the new concrete that was asymmetrically added to the structure. The horizontal deformed shape is presented in Figure 4.33.

The vertical deformed shape was registered during the bridge load test and compared with dial gauges. Load pattern 'A' consisted of six trucks placed on the second span of the bridge and load pattern 'B' consisted of six trucks placed on the third span. The diagram of vertical displacements and comparison with dial gauges is presented for pattern 'A' in Figure 4.23 and for the pattern 'B' in Figure 5.45.

One dial gauge point is not coherent with the calculated vertical displacement diagram. This can be explained by the typical 1 mm manual reading error of dial gauges.

SHM allowed the gathering of precious information concerning the behaviour of the Versoix Bridge. In particular, it was possible to observe the effect of different concrete types on the hindered shrinkage and to detect and observe the appearance of cracks in some of the sections. The horizontal measurements clearly showed a bending induced by the differential shrinkage produced by the asymmetric distribution of the added concrete. The vertical displacements recorded during the load test were in excellent agreement this those obtained with dial gauges and confirmed the good performance of the bridge.

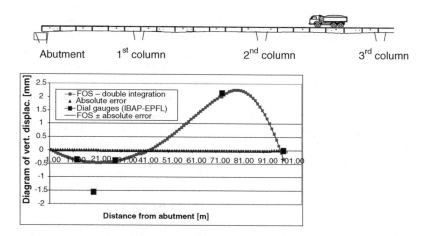

Figure 5.45 Vertical displacement during the load test, pattern 'B', and comparison with dial gauges (Vurpillot, 1999).

5.4.6 Monitoring of a Balanced Cantilever Bridge

A balanced cantilever is an isostatic linear element consisting of two symmetrical cantilevers supported on a column in the middle. The two cantilevers have, by preference, the same weight; thus, the bending moment acting on the supporting column due to the dead load of the cantilevers is balanced. A balanced cantilever bridge is, in general, subject to the same influences as a simple beam: bending, uniaxial or biaxial, shear forces, normal force (prestressed concrete) and torsion (asymmetric load). An example of a balanced cantilever bridge is presented in Figure 5.46.

The cantilever is subject to negative bending moments reaching a maximum in the encast-ered extremity, at the supporting columns. It is shaped according to the bending moment's distribution: the cross-section at the free extremity has a smaller height than that located at the encastered extremity. The cross-section, therefore, is not constant along the girder and its elastic line is not straight (see Figure 5.46).

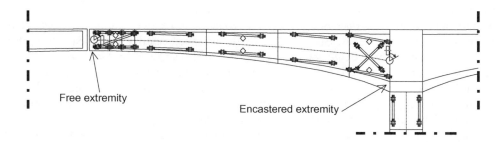

Figure 5.46 View of balanced cantilever bridge with proposed sensors network (courtesy of SMARTEC).

The monitoring strategy for monitoring the deformed shape of a cantilever is similar to that for a simple beam: parallel topologies are to be used and sensors installed parallel to the elastic line of the beam. An unbalanced moment generated by live load is transferred from the cantilevers to the column. To monitor bending caused by the unbalanced moment, it is recommended to install one parallel topology at the top of the column.

A particular issue related to a balanced cantilever bridge is bending and settlement of the free extremity. This issue is often combined with horizontal movement of the free extremity. In order to monitor these phenomena more accurately, the parallel topologies are to be installed more densely – shorter sensors are to be installed in at least three cells. In addition, an inclined sensor, forming a triangular topology with a chain of lower sensors from parallel topologies, can be used to evaluate relative displacement of the free extremity.

The maximum shear force and torsion (if any) occur at the encastered extremity; consequently, crossed topologies are to be installed at that location. Rotations of the supporting column in both vertical planes are to be monitored using two inclinometers installed at the encastered extremity. Two other inclinometers are to be installed at the free extremity in order to monitor rotation in the vertical–longitudinal plane of the free extremity due to bending and rotation in the vertical transverse plane due to torsion. Since the cantilever is an isostatic structure, these two inclinometers also provide for damage detection that may occur at locations not equipped with sensors (e.g. at a crack between two parallel topologies), as in the case of a simple beam (see Section 5.4.2).

Differential settlement of the supporting columns, which may cause significant misalignment of two balanced cantilevers of the same bridge, cannot be measured either with deformation sensors or with inclinometers. Thus, an optical-fibre hydrostatic-levelling system is to be applied. The hydrostatic-levelling system can also be applied to monitor differential settlement of the free extremity.

The discussion concerning the positioning of sensors in the cross-section and the stiffening diaphragm issue of the balanced cantilever bridge is similar to that of continuous girders (see Section 5.4.4). Monitoring of columns and foundations of a balanced cantilever bridge is similar to that presented in Sections 5.2 and 5.3.

An on-site example of monitoring of balanced cantilever bridge is presented next.

5.4.7 On-Site Example of Monitoring of a Balanced Cantilever Girder

The North and South Lutrive Bridges (Switzerland) (Inaudi et al., 1998a; Vurpillot, 1999) are two parallel twin bridges. Each bridge supports two lanes of the Swiss national highway RN9 between Lausanne and Vevey. Built in 1972 by the balanced cantilever method with central articulations, the two bridges are gently curved ($r = 1000$ m) and each bridge is approximately 395 m long with four spans. The two bridges have the same cross-section, which consists of a box girder with variable height (from 2.5 to 8.5 m) and two slightly asymmetric cantilevers, with the purpose of reducing the effects of torsion in the curved bridges. Each balanced cantilever is supported by a pair of slim columns. A view of the Lutrive bridges is shown in Figure 5.47.

Excessive vertical deformation of the bridge occurred, making driving uncomfortable. The supposed source of deformation was creep, and the bridge shape was rectified by post-tensioning. A monitoring system based on a traditional hydrostatic levelling and inclinometers was installed in 1988 and complemented by long-gauge optical-fibre sensors in 1997. The aim

Figure 5.47   View of Lutrive Bridge, Switzerland (Vurpillot, 1999).

of monitoring was evaluation of bridge performance after the rectification and monitoring of its behaviour in the long term.

The west cantilever belonging to the third span of the south bridge, 65.75 m in length, was equipped with deformation sensors installed in the interior of the box girder, on the surface of lateral walls. The cantilever was divided into five cells of 6 m in length with six sensors used in each cell: two sensors for each web and two sensors placed on the upper flag. Six sensors were used for redundancy reasons. Crossed sensors were not used, since torsion and shear strain were estimated not to be an issue. The position of the deformation sensors in the bridge is presented in Figure 5.48.

The deformation sensors could not be installed exactly parallel to the elastic line of the cantilever due to some issues met on site. This generated slight error during the data analysis.

In November 1997, the Reinforced and Pre-stressed Concrete Institute (IBAP) of the Swiss Federal Institute of Technology of Lausanne (EPFL) performed a static load test on the Lutrive highway bridge. The loads were two trucks of 25 tons. Different combinations of loads were tested, depending on truck positions on the bridge.

The deflection (relative vertical displacement) was measured using a hydrostatic-levelling system and inclinometer measurements on five cross-sections equally spaced on the cantilever. The deformation measurement values serve as reference points for the measurement with the deformation fibre-optic sensors.

The rotations measured by the inclinometers were used as boundary conditions in the curvature integration algorithm. The deformed shape and diagram of vertical displacements were calculated and compared with results obtained from the hydrostatic-levelling system. A comparison of the results obtained with the two methods is presented in Figure 5.49 for vertical displacement of the free extremity of the cantilever. The agreement of the two methods is very good, with a maximum error of 6.5 %.

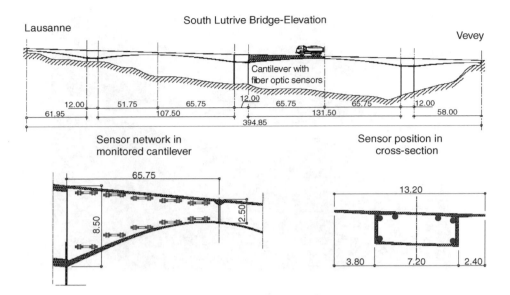

Figure 5.48 Monitored cantilever and positions of deformation fibre optic sensors in Lutrive Bridge (Vurpillot, 1999).

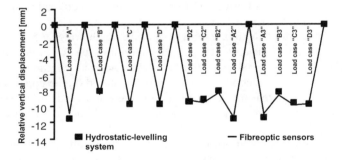

Figure 5.49 Comparison between measurements of free extremity relative displacements obtained with hydrostatic-levelling and fibre optic systems for different load cases (Vurpillot, 1999).

After the load test, measurements were taken over 24 h with a time step of 2 h. Once again, the relative vertical displacement of the free extremity was calculated from optical-fibre sensor measurements and compared with the results obtained from the hydrostatic levelling system. Comparison confirmed good agreement, as shown in Figure 5.50.

Although the network of fibre-optic sensors was aimed mainly at measuring the static short- and long-term spatial displacements resulting from the daily temperature variations and creep, it was found that these same sensors could also be used to capture the quasi-static part of the dynamic deformation of the bridge under traffic load (Inaudi *et al.*, 1998a). The measurement system could acquire measurements only at intervals a few seconds apart. However, it was found that these 'snapshots' could give interesting information concerning the low-

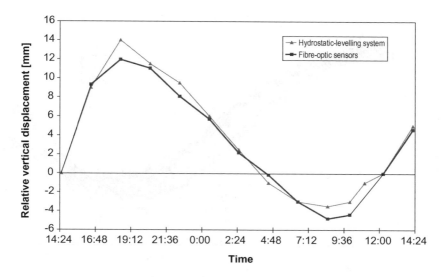

Figure 5.50 Comparison between measurements of free extremity relative displacements obtained with hydrostatic-levelling and fibre optic systems during 24 hours campaign (Vurpillot, 1999).

frequency quasi-static deformations of the bridge. The data from parallel sensors was combined to obtain information about the instantaneous curvature variations of the bridge. These curvature data were than analysed statistically to extract information about the dynamic traffic loads.

In order to test the statistical analysis algorithm, two sensors placed at a quarter span (see Figure 5.48) were measured over 24 h under traffic loading. Both sensors were measured quasi-simultaneously, with a delay of less than 0.1 s; therefore, the measurements are considered to be correlated. The results of the measurements obtained by the two sensors are presented in Figure 5.51.

The measurements of both sensors were scattered around clearly identifiable curves. These curves actually show the deformation generated by a temperature variation of approximately 1 °C. The temperature influence is practically removed after calculation of the curvature. The influences of rheologic effects on curvature can be neglected because the test was performed over just 1 day. Thus, the calculated curvature contained practically only the elastic component. The elastic curvature calculated from the sensor measurements is presented in Figure 5.52.

As expected, most of the values obtained lie around a zero elastic curvature, corresponding to the mean static state of the bridge. High curvature values indicate a deformation induced by the passage of a truck. It can be noticed that these points are concentrated during the day when truck circulation is allowed according to traffic regulation in Switzerland. This is not, however, the case for one reading at around 21:00, since no trucks were registered during the night. Most curvatures are positive and indicate a downward bending of the bridge under the quasi-static loading of the truck. A few points lie outside the noise floor curve for negative curvatures. These are either rebounds of the bridge after a truck leaves the instrumented span or deformations induced by trucks on neighbouring spans.

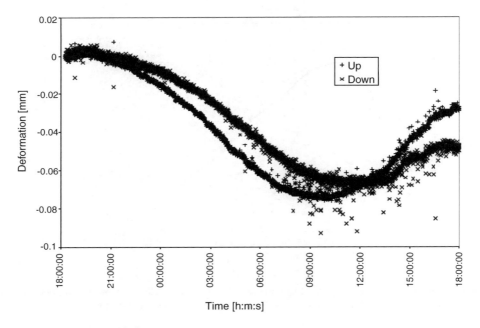

Figure 5.51   Deformation measured by two parallel sensors placed at quarter-span in the Lutrive Bridge (Inaudi *et al.*, 1998b).

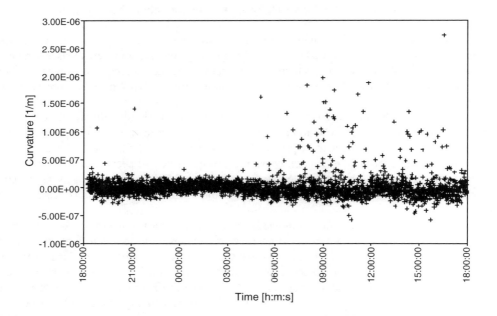

Figure 5.52   Elastic curvature obtained from measurements of parallel sensors (Inaudi *et al.*, 1998b).

If the structure undergoes time-dependent elastic curvature changes $\Delta\kappa_E$, t, then it is possible to associate a probability function $P(\Delta\kappa_E)$ describing the relative probability of finding the structure in a state with elastic curvature of $\Delta\kappa_E$ at any given time. The function $P(\Delta\kappa_E)$ is determined from the results presented in Figure 5.52. The statistical analysis was preformed (1) on the whole set of data ('Day and Night', 3269 data points) and (2) night data only ('Night', 996 points) with Monte Carlo simulation based on simulated deformation measurements with a standard deviation of 1 μm (noise floor, 3267 points). The relative frequency curves for both sets of data are given in Figure 5.53.

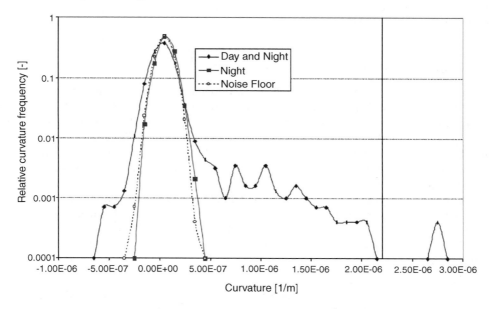

Figure 5.53 Relative curvature frequency (Inaudi *et al.*, 1998b).

The good agreement between the two curves indicates that the noise floor of the fibre-optic system limited the measurements during the night. During the day the higher deformations due to the passage of the trucks are clearly visible. The vertical line in the figure indicates the curvature that was obtained during the previously presented static load test.

Since the load limit on the Swiss highway network was 28 t, the results presented in Figure 5.53 indicate that the dynamic effects on the bridge are very limited. A rebound of a passing truck could, for example, induce higher curvatures than this same truck statically placed on the bridge. The isolated event at 2.7×10^{-6} m^{-1} indicates either an overweight truck (some 40t trucks are also circulating) or two trucks on the bridge at the same time.

This same figure can be used to quantify the probability of having a truck on the bridge at any given time. This is given by the ratio of the number of points of the 'Day and Night' curve inside and outside the 'Noise Floor' curve. This ratio gives 90 % and indicates the probability of obtaining a reliable static deformation value when the measurements are performed under traffic conditions. This shows that it is possible to obtain a reliable static deformation measurement even without stopping the traffic on the bridge. It is sufficient to perform three or more measurements on each sensor and discard any statistically aberrant value to obtain a reliable value. A theoretical analysis considering the truck traffic density, the mean truck speed and the mean transit time gives a slightly lower value of 80 %.

The combined monitoring system, consisting of inclinometers, hydrostatic-levelling system and fibre-optic deformation sensors, provided important information concerning the performance of bridge after the rectification. Relative vertical displacements measured with direct measurements, using the hydrostatic-levelling system, have been in good agreement with those obtained indirectly, from optical-fibre deformation sensor measurements. A quasi-dynamic analysis was performed and demonstrated the capacity of a static system to provide reliable information even under dynamic traffic loads and, moreover, to detect unusual and potentially risky events, such as overloading.

5.4.8 Monitoring of an Arch Bridge

An arch bridge consists of the arch itself, vertical members (columns, struts or ties) and the deck. The load is transferred from the deck to the vertical members, then from the members to the arch, which is the main structural element that transfers the loads to the foundations. The deck consists of simple beams or continuous girders, and the monitoring strategy for these types of structure is presented in Sections 5.4.2 and 5.4.4. The vertical members can be monitored in a similar manner to the columns (see Section 5.3). The monitoring strategy for an arch is presented in this section.

The arch can be isostatic or hyperstatic, depending on the boundary conditions and the presence of hinges. The dominant influences in arches are normal forces, but bending moments, shear forces and torsion are also present. The arch connects significant distances and, consequently, it may have great length. A schematic representation of a concrete arch bridge is given in Figure 5.54.

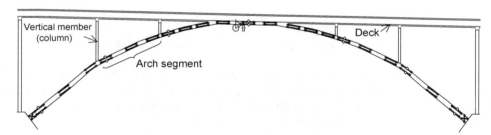

Figure 5.54 Schematic representation of concrete arch bridge and its sensor network (courtesy of SMARTEC).

The particularity of the arch is that the points of application of the live loads are known: these are the points of connections between the vertical members and the arch. Between these points, the only acting force is the dead load, and the segment of the arch between two vertical members can be considered as a curved beam subject to loads only at the extremities. Thus, the member of the arch can be monitored with three parallel topologies: two at the extremities and one in the middle of the segment. Where the segment is not curved too much (e.g. the normal and shear forces at the extremities do not generate a significant moment in the middle of the segment), the parallel topology in the middle of the arch might not be necessary. Temperature sensors are necessary for monitoring thermal strain and its separation from other strain components monitored with deformation sensors.

Besides the common issues that may occur in the other types of bridge, such as differential settlement of foundations or damage that weakens some cross-sections, the particular issue that may occur in an arch is buckling. It can be detected through the monitoring of the deformed shape generated by bending in the case of hyperstatic arches.

In the case of an isostatic arch it is recommended to install at least one inclinometer in the vertical–longitudinal plane in order to detect differential settlements of foundations or damage in a cross-section not equipped with sensors (similar to the simple beam). It is recommended to install an inclinometer even for hyperstatic arches, since it will provide for better accuracy and certain redundancy in monitoring.

The arch deformed shape and diagram of vertical displacements depends not only on bending, but also on axial shortening generated by normal (axial) forces. The arch with only normal forces has zero curvature, but is still deformed (e.g. parabolic arch with a constant cross-section loaded with a dead load). Consequently, both influences must be taken into account when determining the arch deformed shape and diagram of vertical displacements from monitoring. This complete analysis can be very time consuming and complex, since it depends on the geometrical properties of the arch and the number of instrumented cells. However, the monitoring aims can often be achieved with a partial analysis, which consists of determination of the deformed shape and the diagram of radial displacements generated by bending moments only. The issues discussed above that may occur in an arch will be detected by performing the partial analysis.

The shear forces are present at the extremities of the arch segments, but their intensity is usually very low compared with normal forces. Consequently, the monitoring of shear strain using crossed topologies might not be necessary, but should be considered and omitted only after a serious analysis of internal forcess and structural properties of the arch.

Torsion is monitored with crossed topologies installed at the extremities of the arch and an inclinometer measuring rotation in the vertical–transverse plane in the middle of the arch. The typical sensor network for arch monitoring is presented in Figure 5.54.

Monitoring of columns provides redundancy and a good complement to the sensor network presented above. Another good complement is monitoring of the relative vertical displacements in the deck, above the columns, using a hydrostatic-levelling system.

The arch often has a variable cross-section in the form of a box; thus, the discussion concerning the position of sensors in the cross-section and the stiffening diaphragm issue of the arch bridge is similar to that presented for continuous girders in Section 5.4.4. An on-site example of monitoring of an arch bridge is presented next.

5.4.9 On-Site Example of Monitoring of an Arch Bridge

The 'Siggenthal' Bridge is a concrete arch bridge with an arch span of 117 m, built over the Limmat River in Baden, Switzerland (Inaudi *et al.*, 2001). An artistic rendering of the bridge is presented in Figure 5.55.

The arch has a variable width, ranging from 10 m at the extremities, where it doubles in two parallel and distinct arch segments (see Figures 5.58 and 5.59), to 8 m at the top. Its thickness is 0.8 m at the top and 1.4 m at the extremities. The arch curve is made of seven

Figure 5.55 Artist's rendering of the completed bridge (Inaudi *et al.*, 2001).

segments with inflection points under the columns supporting the deck and slightly curved in between.

The deck is a longitudinally and transversally prestressed box girder with constant height. It is supported on the arch by two pairs of columns on each side. In the central section, the arch and the deck fuse into a single structure. The bridge also included two approaches with one span on one side and three spans on the other (see Figure 5.56). The total length of the bridge is 217 m.

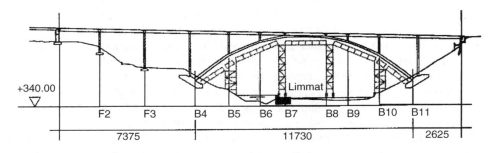

Figure 5.56 Schema of scaffolding during arch construction (Inaudi *et al.*, 2001).

The arch construction proceeded in five successive concrete pouring phases, executed symmetrically and starting from the feet. After lowering the scaffolding, the arch was left free to stand unsupported. The scaffolding during construction is shown schematically in Figure 5.56. During the construction of the vertical columns and of the deck, the arch was stabilized by two temporary towers under the first columns, as shown in Figure 5.57.

Taking into account the different stress and strain states generated in the arch during the different phases of construction and the very low cross-sectional thickness for a span of more than 100 m, it was decided to perform monitoring in order to control the deformation and verify the performance of the structure. The bridge was instrumented with 58 long-gauge fibre-optic deformation sensors, two inclinometers and eight temperature sensors according to the schema

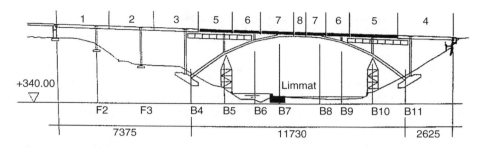

Figure 5.57   Schema of scaffolding during arch construction – second phase (Inaudi *et al.*, 2001).

Figure 5.58   Completed arch of Siggenthal Bridge (Inaudi *et al.*, 2001).

presented in Figures 5.54 and 5.59. The deformation sensors were embedded in the concrete during pouring.

The completed arch is presented in Figure 5.58.

The aims of the monitoring system installed are as follows:

- Monitoring of the concrete deformation

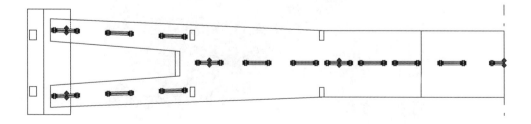

Figure 5.59   Plan view of the sensor network on the left side of the bridge (courtesy of SMARTEC).

- Monitoring of curvatures in the vertical–longitudinal plane
- Determination of the deformed shape due to bending moments.

The monitoring is carried out during the whole lifespan of the bridge, but with particular interest during the following phases: pouring of concrete of the different arch sections, removal of the scaffolding, free-standing phase of the arch, installation of the temporary towers, construction of the supporting columns and the deck, bridge testing, long-term and in-service performance. The first three phases are presented in this section.

The horizontal bending and torsion were estimated as very low and in order to decrease the costs were not monitored. The parallel topologies of the sensors consisted of only two sensors installed on the vertical principal axis of the cross-section. Each segment of the doubled arch extremity was equipped as a separate segment. The plan view of the sensor network on the left side of the bridge is presented in Figure 5.59. The sensor network on the right side of the bridge was symmetric with the network on the left side of the bridge.

The connecting cables of all the sensors were led to the temporary central measurement point installed near to one abutment. The central measurement point contained the reading unit (SOFO type) and channel switches and is shown in Figure 5.60.

The deformations measured by three pairs of sensors installed in one of the arch extremity segments are presented in Figure 5.61. In the first phase (before 10 July 2000) all sensors measure a similar shrinkage. On 10 July 2000 the second section of concrete was poured on top. Because of the slightly curved shape of the arch, this additional load induces a bending in the section that is visible in the figure: the sensors installed in the upper part of the cross-section (denoted 'BO') measure an elongation, whereas the sensors installed in the lower part of the cross-section (denoted 'BU') measure a shortening. After that event the sensors restart to measure the same shrinkage behaviour (the curves in Figure 5.61 become approximately parallel).

The typical average curvature measured in a cell of the arch during scaffolding removal and free standing of the arch is given in Figure 5.62. The measurements performed before 13 August 2000 show an irregular behaviour corresponding to the different phases of the scaffolding removal. As soon as the arch is completely free, a periodic behaviour, caused by

Figure 5.60   View of temporary central measurement point (courtesy of SMARTEC).

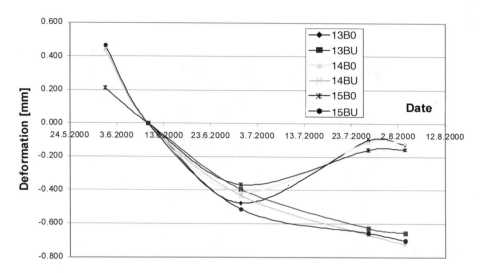

Figure 5.61   Concrete deformations in the segment at the extremity of the arch during construction (Inaudi *et al.*, 2001).

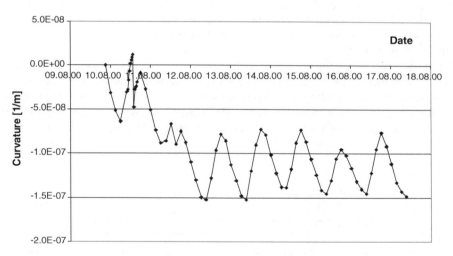

Figure 5.62   Typical average curvature measured in an arch cell during and after removal of scaffolding (Inaudi *et al.*, 2001).

daily temperature variations, was observed. It is interesting to note that the curvatures induced by the dead-load activation are of the same order of magnitude as those induced by the daily temperature variations and sun irradiation (on a late-summer day).

The deformed shape and radial displacements generated by bending moments are obtained using double integration of the curvature distribution. The zero displacements of the abutments were used as boundary conditions. A diagram of the radial displacements (deformed shape) with respect to the curvilinear elastic line of the arch (abscissa) generated by bending moments

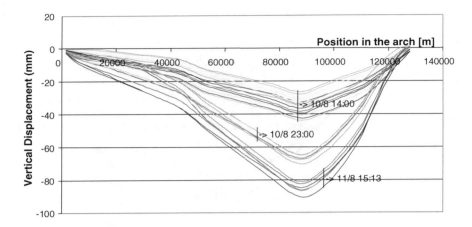

Figure 5.63   Radial displacement generated by bending during the removal of scaffolding (Inaudi *et al.*, 2001).

during the removal of scaffolding is presented in Figure 5.63. Each curve was taken at 30 min intervals. As expected, the arch moves inwards when the scaffolding is removed. The scaffolding is first lowered on the left side, then on the right. These operations were performed stepwise, which is also observable from the measurements. Note that the deformation is not completely symmetrical.

The radial direction for the point near to the mid-span coincides with the vertical direction, and the displacement in the radial direction is actually a vertical displacement. The vertical displacement generated by bending moments near to the mid-span during scaffolding removal and during the arch free-standing phase is presented in Figure 5.64. Again, it is possible to observe the different lowering phases and the periodic movement due to daily temperature variations.

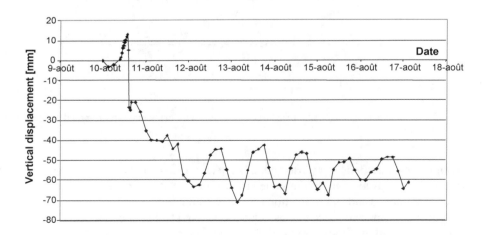

Figure 5.64   Vertical displacement near to mid-span during and after the removal of scaffolding (Inaudi *et al.*, 2001).

The total vertical displacement was about 100 mm from direct measurements done through triangulation, compared with approximately 60 mm determined by the deformation sensors. The difference of approximately 40 mm between the two measurements can be explained by the contribution of normal force (axial strain) to the vertical displacement determined by triangulation, but not calculated by the double-integration algorithm. It is once more interesting to note that the radial displacements due to the activation of the dead load are of the same order of magnitude as the daily movements due to sunshine.

An optical-fibre monitoring system was applied to the Siggenthal arch bridge. The sensors were installed successfully and the arch was monitored during the different construction phases. It was possible to measure the local deformations of the concrete and to determine the radial displacements during scaffolding removal and during the free-standing period of the arch. Therefore, it was possible to observe the deformation of the arch when being loaded by its dead load and by daily temperature fluctuations. It was found that during summer the daily temperature influence on the arch is particularly large. Measurements during the other phases of construction were also performed, but taking into account the results presented in previous sections, and in order to keep this section concise, these measurements are not presented.

5.4.10 Monitoring of a Cable-Stayed Bridge

A cable-stayed bridge is a complex structure consisting of three main members: deck, pylon and cables. Monitoring of these main members will be the focus of this section. Monitoring of the other members, namely columns and foundations, were presented in Sections 5.2 and 5.3.

Cable-stayed bridges are commonly very long, with spans as long as several hundreds of metres. Owing to their large dimensions, a large number of sensors may be required. In this section, a monitoring strategy oriented to a relatively limited number of sensors is developed leading to an economically affordable solution. The proposed sensor network is shown schematically in Figure 5.65.

The influences of bending and normal forces in the first span of the deck (see Figure 5.65) can be monitored with four to six parallel topologies. Cell 1 is positioned as close as possible to the pylon. The maximum bending moments and normal forces of the bridge are expected in this cell. Cell 3 is to be placed close to cable anchoring zone. This cell will capture the influences of the bending moment and maximum normal force introduced directly from the cables and help detect relaxation or damage in the cables. Cell 2 is positioned between cell 1 and cell 3 (middle distance) in order to better interpolate the deformed shape of the deck. In this cell the bending moment may be important due to live loads. Cell 4 is installed in the middle of the cable anchoring zone with the aim to detect relaxation or damage in the cables installed to the left of the cell. Cell 5 is installed on the other side of the cable anchoring zone with the aim to monitor the maximum bending influences in that part of the deck non compressed with cables. Cell 6 is installed close to the abutment in order to determine the boundary conditions precisely. However, if the deck may not prestressed, then it may not be necessary to monitor this cell, since neither bending nor axial forces are present in that case.

The sensor network for the second span is similar to that for the first span. The difference is that the part of the span between the two cable anchoring zones (cells 5, 6 and on) is to be monitored as a span of continuous girder (minimum three cells, recommended five cells; see Section 5.4.4). Differential settlements of foundations and torsion are detected with two

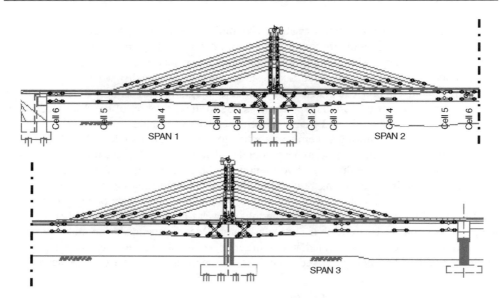

Figure 5.65 Schematic representation of monitoring strategy applied on a cable-stayed bridge (courtesy of SMARTEC).

inclinometers measuring in two perpendicular vertical planes installed in the middle of the second span. Depending on length of deck more inclinometers can be installes along the deck. Temperature sensors are installed at regular intervals. Finally, the third span is equipped with a similar sensor network to that used for the first span. The cross-section of the cable-stayed bridge is commonly very wide and it might be necessary to install more than four sensors in the cross-section. Possible positions of sensors in the cross-section are presented in Figure 5.66 (see also the example in Section 5.4.7).

The crossed topologies installed in cell 1 of each span are used for monitoring the shear strain generated by the shear force and torsion. For a very wide cross-section the torsion can be evaluated indirectly, by separate analysis of deformed shapes determined from pairs of parallel sensors belonging to the same vertical plane (e.g. the network of parallel sensors installed on the left side of the cross-section measures a different deformed shape from the network of parallel sensors installed on the right side of the cross-section). If the torsion is estimated not to be an issue, then an indirect evaluation gives satisfactory results and the crossed topologies can be omitted.

The stay cables consist of bundles of strands. The strain in the cable is monitored directly, by a sensor (simple topology) installed on the cable strand. The sensor directly detects a malfunction in the cable on which it is installed.

The best performance is obtained if each cable of the bridge is monitored. For bridges with a large number of cables, every second cable can be monitored in order to decrease the costs. Failure of non-monitored cables is detected indirectly through an increase in strain in the monitored cables. The sensors installed in the deck also detect malfunctions of the cables (the deformed shape of the deck changes). The sensors installed in cell 4 help localization of the problematic cable. Temperature sensors are to be installed in parallel with deformation sensors.

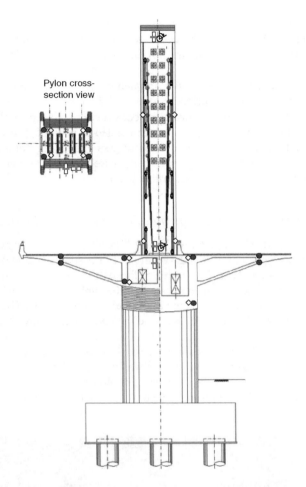

Pylon cross-
section view

Figure 5.66 Schematic representation of monitoring strategy applied to a pylon of a cable-stayed bridge; the positions of sensors in the deck cross-section are also indicated (courtesy of SMARTEC).

The pylon is to be instrumented using a combination of the solution for the deck and a cantilever (see Section 5.4.6). The lowest cell is to be as close as possible to the deck. This cell is subject to maximum normal forces and bending. The second cell is to be as close as possible to the cable anchoring zone on the pylon, but out of this zone. This cell is directly loaded by the resultant force from the cables. There is no perpendicular load between these two cells; consequently, there is no need to install sensors in between. The third cell is to be set in the middle of the anchoring zone; finally, the fourth cell is at the end of the cable anchoring zone.

The rotations in both vertical planes are to be monitored with inclinometers set on the top of the pylon. In order to increase the performance of the monitoring system and provide for a certain redundancy, it is also recommended to install inclinometers at the bottom of the pylon. Temperature sensors are very important, due to height of the pylon and its exposure to the sun.

Analysis of the monitoring results is enhanced if additional sensors are installed: fibre-optic accelerometers in the deck, on the cables and at the base and at the top of the pylon; an anemometer on the top of the pylon; an absolute movement monitoring system (e.g. GPS based) on the top of the pylon.

The construction of a cable-stayed bridge frequently starts from the pylons, and then the deck is built on both sides. A long bridge makes it difficult and expensive to centralize the system. In order to allow monitoring during the construction and reduce the costs of sensor cable installation and interconnection, it is recommended to install two central measurement points in the bases of two pylons. Two central measurement points can communicate with each other as well as with a remote monitoring centre using wired or wireless connection. An on-site example of monitoring of the cable-stayed bridge is presented next.

5.4.11 On-Site Example of Monitoring of a Cable-Stayed Bridge

The new cable-stayed bridge built in Port of Venice 'Marghera', at the inner edge of the Venice lagoon, consists of two spans of 105 and 126 m (Del Grosso *et al.*, 2005). The deck of the bridge has a circular-shaped elastic line with a radius of 175 m. It is built of hybrid steel and reinforced concrete and supported on each side by nine cables. The cables are attached to an approximately 80 m tall and slightly inclined reinforced concrete pylon. The bridge represents a new road link between the Venice port areas and the national highway. A view of the bridge is shown in Figure 5.67.

The importance of the bridge and its complex curved structure were the reasons to implement a system for long-term monitoring. The system was installed during the construction of the bridge in order to monitor all important construction phases. The aim of the system was to perform both continuous static monitoring and periodic dynamic monitoring. In order to select

Figure 5.67   View of cable-stayed bridge in the Port of Marghera (Del Grosso *et al.*, 2005).

the locations of the sensors, a dynamic analysis of the bridge was performed using the FEM. The periodic dynamic monitoring was estimated as necessary in order to perform identification of structural characteristics.

The deck of the bridge consists of two lateral 'I'-shaped steel girders and a central box steel girder, mutually connected by transverse diaphragms. The prefabricated concrete slabs were laid on the steel girders and fixed with a cast-on-site layer. The FEM analysis helped to identify four sections in each span (eight sections in total) that were to be equipped with parallel topologies. Each topology consisted of six sensors installed on both lateral girders and the web of the central girder. The gauge length of the selected sensors was 1.5 m.

The rotations were monitored at the ends of the spans supported by cable using two optical-fibre inclinometers. Temperature sensors were installed in two intermediate cells of each span. In total, 48 deformation sensors, 2 inclinometers and 24 temperature sensors were installed in the deck. Views of the sensors and inclinometer installed on the steel girders by surface mounting are presented in Figures 5.68 and 5.69 respectively.

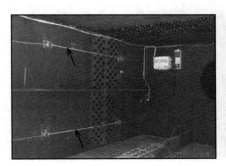 

Figure 5.68   View of sensors installed on central box girder (left) and lateral 'I' girder (right) (courtesy of Tecniter).

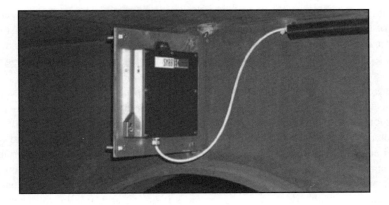

Figure 5.69   View of optical-fibre inclinometer installed at extremity of central box girder (courtesy of Tecniter).

Figure 5.70 View of pylon under construction (left) and positioning of the sensor attached to the rebar (right) (Del Grosso *et al.*, 2005).

The pylon is built of reinforced and partially prestressed concrete. It has a triangular cross-section with decreasing dimensions towards the top. The pylon is not vertical, but inclined for angle matching to the direction of the resultant force from the cables. The weight of the pylon was balanced by post-tensioning; thus, the most important efforts are normal forces, whereas bending occurs only due to wind and temperature variations.

Four cells were selected along the pylon and equipped with parallel topologies. Each parallel topology contained three 7.5 m long sensors. Each sensor belonging to the parallel topology was installed in one corner of the triangular cross-section by embedding.

Two mutually perpendicular fibre-optic inclinometers and one anemometer were installed at the top of the pylon. The two upper cells were equipped with temperature sensors. In total, twelve deformation sensors, six temperature sensors, two inclinometers and one anemometer were installed on the pylon. A view of the pylon under construction and positioning of a sensor attached to a rebar is presented in Figure 5.70.

There are nine cables in total on each side of the pylon, with 31–85 strands. In order to monitor the cables' performance it was decided to install a 250 mm deformation sensor on one strand of each cable. These sensors monitor the strain in the strands and are supposed to detect unusual behaviour caused by damage. Based on the results obtained from monitoring, it can be decided whether to perform inspections using the magnetic flux leakage method. In addition, during dynamic test sessions, some accelerometers may be installed temporarily on the cables.

The sensors were installed by clamping and gluing. First, the protection coating of the cable was cut and removed, then the sensor was installed and, finally, the protection coating was reinstalled. Views of deformation sensors installed on the cable strands are given in Figure 5.71.

Figure 5.71   Views of sensor attached to strand of the cable (Del Grosso *et al.*, 2005).

All the connecting cables are connected to the central measurement point installed in the base of the pylon. The static reading unit is permanently connected to sensors and performs long-term static readings of all the sensors. The dynamic reading unit can be manually attached to the same sensor network when needed. Remote control of the system is established via a wired telephone line. A view of the central measurement point is given in Figure 5.72.

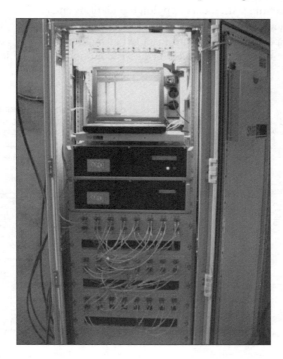

Figure 5.72   View of central measurement point installed in the base of the pylon (courtesy of Tecniter).

The fibre-optic monitoring system applied on the Port of Marghera cable-stayed curved bridge is presented in this section. The sensor network was built in accordance with the principles described in Section 5.4.6. The results of monitoring are not disclosed, which is the reason they are not presented here.

5.4.12 Monitoring of a Suspended Bridge

A suspended bridge, with a span exceeding 1 km, is one of the most impressive man-made structures. It is used to connect very distant points and the dimensions of its members are very big; therefore, a large number of sensors may be required for monitoring of a suspended bridge. The main members of the suspended bridge are the deck, pylon, suspending cable and vertical cables. Similar to a cable-stayed bridge, a monitoring strategy dealing with a limited number of sensors and oriented to the main structural members is presented in this section. The proposed sensor network is shown schematically in Figure 5.73.

A segment of the deck between two vertical cables is subject to a linear or near-linear distribution of bending moments. The extreme values of the moment occur at the extremity of the segment (i.e. in the cross-section close to the vertical cable). Therefore, it is sufficient to equip with parallel topology only the extremity of the segment where the maximum absolute moments occur, as shown in Figure 5.73. Only those segments in which the maximum absolute moments for the entire deck occur are to be equipped with two parallel topologies (see cells 10a and 10b in Figure 5.73).

Detection of differential settlements of pylon foundations and torsion is performed with two inclinometers installed at several locations along the span, measuring in two perpendicular vertical planes. Temperature sensors are installed at regular intervals. As for a cable-stayed bridge, the cross-section of the suspended bridge is commonly very wide and more than four

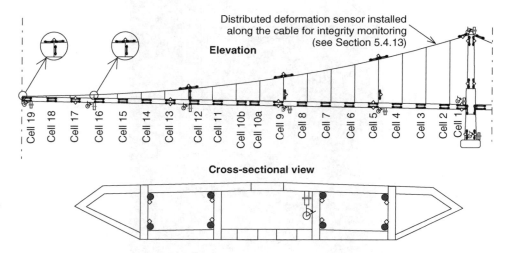

Figure 5.73 Schematic representation of monitoring strategy applied on suspended bridge (only a half span presented) (courtesy of SMARTEC).

sensors are to be installed in the cross-section. Possible positions for sensors in the cross-section are presented in Figure 5.73 (see also the example in Sections 5.4.6 and 5.4.7 concerning the cable-stayed bridge).

The deck of a suspended bridge is frequently made of steel, making calculation of the normal force and bending moments from the monitored strain and temperature relatively simple. Consequently, the shear force distribution along the bridge can be estimated indirectly as the first derivative of the bending moment distribution. The shear force distribution along the deck allows estimation of the normal force in each vertical cable. In addition, as discussed in Section 5.4.10, torsion can be evaluated indirectly for very wide cross-sections by separate analysis of deformed shapes determined from pairs of parallel sensors belonging to the same vertical plane. Consequently, crossed topologies for monitoring shear strain are not necessary.

The horizontal component of the force in a suspension cable is constant along the cable; this fact makes possible installation of only 'control' sensors in the form of a simple topology. Two sensors are installed at the control locations, left and right from the vertical cable link point (node). The vertical cable is also equipped with a simple topology. This sensor configuration makes possible the determination of the horizontal force in the cable, which when combined with the known geometry of the cable makes it possible to evaluate the normal force along the cable. The temperature sensors are to be installed at the node and at the sensor location in the vertical cable.

The position of the sensor along the vertical cable is not of importance, but for practical reasons, it is recommended to install it at a lower level, close to the deck. At a lower location the sensor is easier to access. General considerations related to the installation of the sensor on the cable (cable strand) are presented in Section 5.4.10.

A malfunction in a vertical cable, which is not equipped with sensors, is indirectly detected by the change of strain field in the deck. Sensing damage (reduction of cross-section) in a suspension cable at locations not equipped with sensors is more difficult. The breakage of some strands in a suspension cable has, as a consequence, a local strain change that does not significantly influence the force distribution along the cable and in vertical cables, and creates only a small change in the strain field of the deck. Therefore, it might be difficult to distinguish the difference between a real strain change in the bridge due to damage in a suspension cable and the noise generated by the resolution of the monitoring system.

In order to enhance the monitoring performance of a suspension cable, it is recommended to install a distributed deformation sensor along the cable. The distributed sensor, since it is in continuous contact with the cable, is capable of detecting local strain changes at any point on the cable. In this manner, an integrity monitoring of the cable is performed. More details concerning the integrity monitoring of bridge members is given in Section 5.4.13.

The pylons are equipped with sensors following a strategy similar to that for the cable-stayed bridge. The dominant influences are the normal force from cables and bending moments generated by cable force imbalance, temperature variations and dynamic loads (wind and earthquake). The most loaded cross-section of the pylon is the lowest one; therefore, one cell with parallel topology is to be installed at the base of the pylon (see Figure 5.73). The other important cross-section to be monitored is the one just below the cables, and one more cell is to be equipped with sensors there (see Figure 5.73). Between these two cells the bending moment line is linear, but the cross-section may be variable. That is why it is recommended to equip cells with parallel topologies where the cross-sectional properties of the pylon change (see Figure 5.73).

The rotations in both vertical planes are to be monitored with inclinometers set on the top of the pylon. In order to increase the performance of monitoring system and provide for certain redundancy, it is also recommended to install inclinometers at the base of the pylon. Temperature sensors are very important due to the height of the pylon and its exposure to sun.

Similar to the discussion for cable-stayed bridges, the analysis of monitoring results is enhanced if additional sensors are installed: fibre-optic accelerometers in the deck, on the suspension cable and at the base and at the top of the pylon; an anemometer on the top of the pylons; an absolute movement monitoring system on the top of the pylons (e.g. GPS-based system).

The construction of the suspended bridge frequently starts from the pylons, then the cable is installed and finally the deck is built. A long bridge makes it difficult and expensive to centralize the system. In order to allow monitoring during the construction and to reduce the costs of sensor cable installation and interconnection, it is recommended to install two central measurement points in the bases of two pylons. Two central measurement points can communicate with each other as well as with a remote monitoring centre using wired or wireless connection.

Fiber optic sensors were employed to dynamically monitor the main cable of the suspended bridge L'lle d'Orleans in Canada. At the moment of writting the book, the information concerning the project was not available, but can now be found in the literature (Talbot *et al.*, 2007).

The concept of integrity monitoring of long bridges using distributed monitoring systems with an on-site example is presented in Sections 5.4.9 and 5.4.10.

5.4.13 Bridge Integrity Monitoring

Distributed optical-fibre sensing technology based on the Brillouin scattering effect has opened up new possibilities in structural monitoring. A distributed deformation sensor is sensitive at each point of its length to strain changes and cracks. Such a sensor can be installed over the

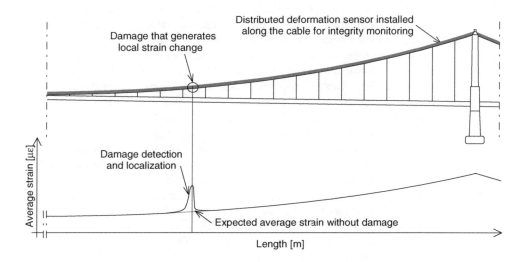

Figure 5.74   Schematic representation of integrity monitoring.

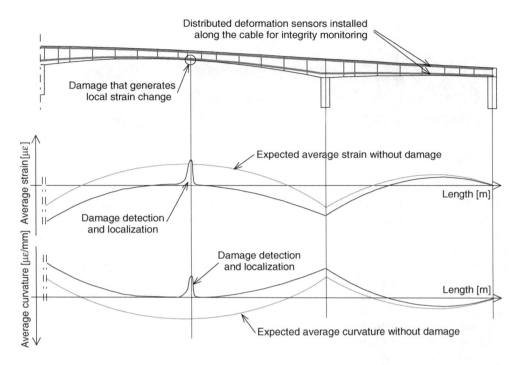

Figure 5.75   Schematic representation of simultaneous integrity monitoring and curvature monitoring.

whole length of the monitored structural member of the bridge, and therefore provide for direct detection and localization of local strain changes generated by damage. The principle of integrity monitoring is represented schematically in Figures 5.74 and 5.75.

Since distributed sensors provide for average strain measurement similar to long-gauge sensors, they can be used with twofold purpose: for integrity monitoring as, presented above, and for curvature and deformed shape monitoring. Two parallel distributed sensors combined in parallel topologies provide a large number of densely distributed cells, making double integration of curvature more accurate. An example is presented in Figure 5.75.

Commercially available distributed monitoring systems are expensive; their use for monitoring of bridges is only justified in cases where a large part of structure is to be monitored or an important impact on society is generated by structural malfunction, or simply where no other technology can fulfil the aims of monitoring. Typical bridge structural members that are candidates for integrity monitoring are the long beams, girders, decks and suspension cables.

An on-site example of bridge integrity monitoring is presented next.

5.4.14 On-Site Example of Bridge Integrity Monitoring

Götaälvbron, the bridge over the Göta River, was built in the 1930s and is now more than 70 years old. Being one of the three communication lines that connect the two sides of the Göta River, Götaälvbron is a bridge of high importance for the city of Gothenburg (Sweden). The

Figure 5.76   View of the nearly 1 km long Götaälvbron Bridge (courtesy of SMARTEC).

bridge is more than 1000 m long and consists of a concrete slab poured on nine steel continuous girders supported on more than 50 columns. During the last maintenance work, a number of cracks were found in steel girders, notably in zones above columns where significant negative bending moments are present. These cracks are consequences of fatigue over the many years of service and mediocre quality of the steel. A view of the bridge is presented in Figure 5.76.

The bridge is now repaired and the bridge authorities would like to keep it in service for the next 15 years, but new cracks due to fatigue can occur again. These new cracks can lead to the failure of cracked girders, which may occur suddenly since damage is generated by fatigue. That was the reason to perform continuous bridge integrity monitoring (Glišić *et al.*, 2007).

The monitoring system selected for this project must provide for both crack detection and localization and for strain monitoring. Since the cracks can occur at any point or in any girder, the monitoring system should cover the full length of the bridge. These criteria have led the bridge authorities to choose a truly distributed fibre-optic monitoring system based on stimulated Brillouin scattering.

Summarizing, the monitoring aim has been to perform long-term integrity monitoring of the bridge. Split into single tasks, the following specifications of the monitoring system have been requested:

1. To detect and localize new cracks that may occur due to fatigue
2. To detect unusual short-term and long-term strain changes
3. To detect cracks and unusual strain changes over the full length of five girders, in total 5 km
4. To perform one measurement session every 2 h
5. To perform self-monitoring; that is, to detect malfunctioning of the monitoring system itself
6. To allow user-friendly and understandable data visualization
7. To send warning messages automatically to responsible entities
8. To function properly for 15 years.

Being the world's first bridge application of a distributed sensing system on such a large scale, the requested specifications imposed a number of challenges to the reading unit performance, deformation sensor production and installation, and data management.

The sensor quality has to guarantee good transfer of strain from the structure to the optical fibre, good mechanical resistance to installation and handling actions, and moderate optical losses. That is why the production parameters were optimized, and every metre of distributed deformation sensor was controlled optically, mechanically and visually during the fabrication.

Figure 5.77 On-site test of distributed sensor gluing procedure (left) and installed sensor for full performance tests (right) (courtesy of SMARTEC).

In order for the system to be able to detect cracks at every point, it was decided to glue the sensor to the steel girder. The crack should not damage the sensor, but lead to its delamination from the bridge (otherwise the sensor would also be damaged and would need to be repaired). A gluing procedure, therefore, was established and rigorously tested in the laboratory and on site. A photograph of the on-site gluing test is presented in Figure 5.77. The full performance was also tested in the laboratory and on site, and a photograph of tested sensors installed on the bridge is also presented in Figure 5.77.

The installation of sensors was a challenge itself. Good treatment of surfaces was necessary and a number of transverse girders had to be crossed. Difficult access and limited working space in the form of a lift basket, often combined with a cold and windy environment and sometimes with night work, made the installation particularly difficult. A view of the bridge girders is given in Figure 5.78.

The distributed deformation sensors were produced in lengths of 90 m. Owing to optical losses, a maximum of three sensors could be enchained, making a total effective length of 270 m. The sensor measurements are compensated for temperature by using a temperature-sensing cable that also has the function of bringing back the optical signal to the reading unit designed for loop configuration. The three enchained distributed sensors and temperature-sensing cable create a basic loop of the system. The basic loop is presented in Figure 5.79.

There are 20 basic loops, two sets of five loops at south part of the bridge and two sets of five loops in the north part of the bridge. Each part of the bridge is monitored by a different reading unit provided with 10 channels. The measurement time for a single loop is less than 10 min for the strain range of 15 000 $\mu\varepsilon$ (-5000 to $+10\,000$ $\mu\varepsilon$) and spatial resolution of 1m. Therefore, the measurement time for all the sensors is less than 2 h.

Figure 5.78   View of bridge girders and part of the installation team (courtesy of SMARTEC).

The physical point A on the bridge is measured twice: first, the temperature is measured at loop coordinate A_T and then the noncompensated strain at loop coordinate A_s. Temperature compensation is performed automatically by the software.

The span is a segment of the girder that is delimited by cross-sections supported by columns. One sensing loop can 'cover' several spans, and two sensing loops can be overlapped within the same span. The general positions of the spans and loops are shown in Figure 5.80.

Event localization is performed with respect to the bridge coordinate system, indicating on which girder it happens (total of five girders equipped) and on which metre of the girder span (total of 55 spans of approximately 20 m). The localization of events in this manner is very efficient and intuitive for the bridge maintenance team.

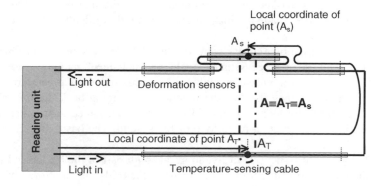

Figure 5.79   Basic sensing loop consisting of three 90 m SMARTape sensors and temperature-sensing cable. Points A_T and A_s are at the same physical location; they are only separated schematically in the drawing in order to simplify presentation (courtesy of SMARTEC).

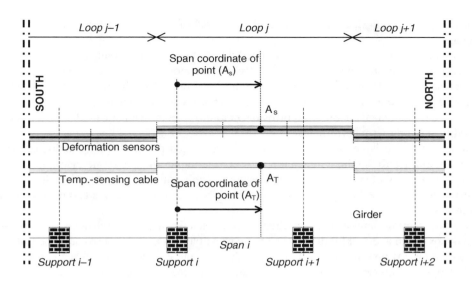

Figure 5.80   General position of spans and loops (Glišić *et al.*, 2007).

The monitoring system is provided with self-monitoring functions: automatic unblocking, detection of different types of malfunction, such as a cut of electrical power, cut of the communication lines, unfavourable environmental conditions (high or low temperature), physical damage of sensors, and so on.

The measurements were performed with a distance sampling interval of 0.1 m, making the total number of monitoring points bigger than approximately 100 000 – 50 000 for noncompensated strain and another approximately 50 000 for temperature. Special software for data handling, crack detection and short-term and long-term strain analysis has been developed. The same software controls functioning of the reading units. In the case of crack detection, detection of unusual strain variation or detection of a malfunction in the system, warnings are sent to responsible entities in the form of an e-mail, SMS and voice message, providing for redundancy in communication. Upon receipt of messages, the responsible entities are supposed to aknowledge they received the messages and to proceed according to established procedures.

A truly distributed fibre-optic monitoring system, based on the stimulated Brillouin effect, is for the first time applied on a large scale for integrity monitoring of a bridge. Five girders, with a total length of approximately 5 km (each girder is approximately 1 km long) are equipped with sensors. The system provided for bridge integrity monitoring, and strain and temperature monitoring in the long term.

5.5 Monitoring of Dams

5.5.1 Introduction

Dams are certainly well-monitored structures. Monitoring of dams is regulated by codes in most countries. This is comprehensible, taking into account the consequences that a dam failure can cause in terms of injuries and lost human lives, social impact, economic losses and ecologic damage.

Dam monitoring strategies were developed during the twentieth century, mainly based on conventional measurement instruments. Therefore, the monitoring strategies based on the use of optical-fibre technologies are similar to traditional ones. The discrete traditional sensors can be exchanged with optical-fibre sensors, which bring advantages in terms of insensitivity to harsh environmental conditions (corrosion, humidity, electromagnetic fields, ...), accuracy, stability and long-term performance.

Since dams are large structures, the use of distributed optical-fibre technologies brings new possibilities for strain, temperature and seepage monitoring. A single distributed sensor allows monitoring of large areas, which was not possible with traditional sensors. In addition, distributed sensing technology provides for integrity monitoring of dams and dam structural members.

By construction, dams can be classified in three general categories: arch dams, gravity dams and dykes (earth and rockfill dams). The basic principles for each of these categories are presented below. At present, optical-fibre technologies cannot completely replace conventional technologies. Therefore, the monitoring strategies presented will be focused only on parameters that can be monitored with optical-fibre technologies. That is the reason why the on-site examples presented in this section are related to monitoring single parameters rather than to monitoring of the dam as a whole.

5.5.2 Monitoring of an Arch Dam

Arch dams are large structures commonly built of nonreinforced concrete. They have the form of a shell curved in the horizontal and vertical planes, in the shape of an arch. An arch dam is commonly built in a mountainous environment and closes the gorge, leading to a significant water accumulation behind.

The arch dam is conceived to accept only compressive stresses (nonreinforced concrete); therefore, one of the main parameters to monitor is strain, which can indicate the occurrence of tensional stresses and cracks. The tensions are mainly generated by water level changes (uplift pressure) and dam deformations, temperature gradients generated during the construction (hydration of concrete) or by environmental seasonal and daily temperature variations, and an alkali reaction in concrete aggregate. The other issue is related to seepage, which can occur in the foundations of a dam or through cracks. Thus, the following monitoring parameters are of particular interest:

1. Strain
2. Deformation
3. Displacement
4. Inclinations
5. Crack detection (integrity)
6. Crack opening (for detected and opened cracks)
7. Temperature
8. Pore pressure
9. Seepage detection.

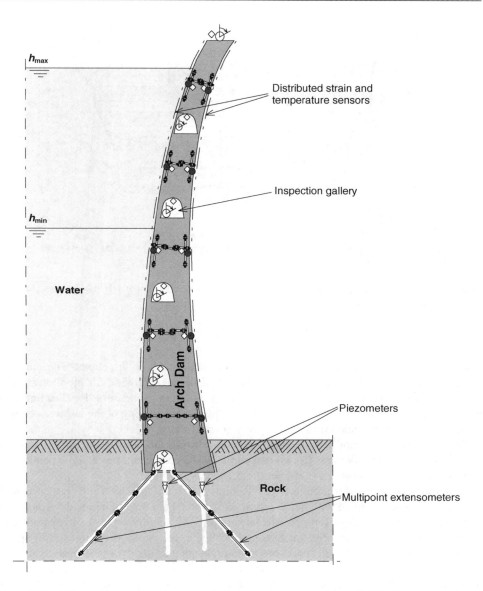

Figure 5.81   Schematic representation of optical-fibre sensor network installed in the cross-section of an arch dam (courtesy of SMARTEC).

The monitoring strategy comprised of the parameters presented above is shown in Figures 5.81 and 5.82. Multipoint extensometers are created by enchaining several single deformation sensors. A multipoint extensometer is to be anchored in the nondeformed layer of soil and provides for absolute displacement of the dam base and strain distribution in the soil.

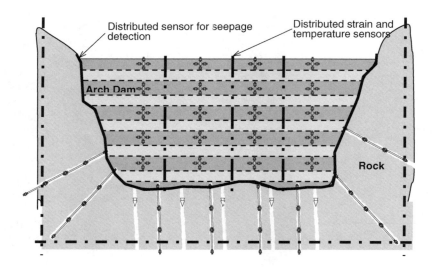

Figure 5.82 Schematic representation of optical-fibre sensor network (partial) installed in an arch dam; view from the downstream side (courtesy of SMARTEC).

Long-gauge deformation sensors can be installed in all three directions. The enchained simple topologies installed through the thickness of the arch provide for monitoring of the average strain distribution, which decreases along the chain of sensors. Parallel topologies installed in the horizontal and vertical planes provide for monitoring of the average strain and curvature distributions, and evaluation of the deformed shape.

Deformed shape cannot be calculated accurately using the double-integration algorithm since, for shells, the curvatures in two perpendicular directions are mutually dependent (the dependence involves Poisson's coefficient).

Inclinometers provide for monitoring of absolute rotation in galleries and on the top of the crown. Inclinometers along with deformation sensors can help determine the deformed shape in the dam more accurately. Temperature sensors provide monitoring of the temperature distribution and thermal strain. Piezometers are installed in the boreholes in the soil in order to monitor pore pressure. Opened cracks are monitored using triangular topologies, as shown in Section 4.5.3. For dams situated in seismic areas it is also important to install accelerometers in inspection galleries. The number and positions of sensors depend on the project requirements.

In order to perform integrity monitoring (see also Section 5.4.13), distributed strain and temperature sensors can be installed along the lines where the maximum tensional stresses are expected. These sensors provide for crack detection, but also for average strain and average curvature monitoring, and can replace discrete deformation and temperature sensors. Distributed temperature sensors can provide for seepage detection. Finally, a separate seepage detection sensor can be installed on the downstream side of the dam, along the boundary with rock (foundations). On-site examples on monitoring of parameters in an arch dam are presented next.

5.5.3 On-Site Examples on Monitoring of an Arch Dam

The first on-site application is related to the pioneering application of an optical-fibre interferometric extensometer at the Emosson Dam (Glišić *et al.*, 1999). The Emosson Dam is situated in the Swiss Alps, 1930 m above sea level (a.s.l.), near the Swiss town of Martigny. Completed in 1975, the dam is 180 m high and its crest is 554 m long; its thickness varies from 9 m (crest) to 48.5 m (footing). A view of the dam is given in Figure 5.83.

Figure 5.83   View of Emosson Dam and optical-fibre extensometer on the spool (Glišić, 2000).

Water is collected during the warm seasons, when snow and glaciers melt, and used during the cold seasons, when the demand for electricity is greater. The dam's maximum capacity is approximately 225×10^6 m^3. There are three principal collectors that guide the water from the Alpine glaciers to the dam: the west and south collectors guide water from the French Alps and the east collector guides water from the Swiss Alps. Two power plants exploit the water: the Vallorcine in France (1125 m a.s.l.) and La Bâtiaz (462.5 m a.s.l.) in Switzerland.

A single point optical-fibre extensometer, named F3, was built to replace the rod extensometer at the Emosson Dam in Switzerland. The geometrical and mechanical properties of the sensor's packaging were imposed by on-site conditions: the external diameter of the packaging was limited by the dimensions of the borehole in which the sensor had to be placed, and the stiffness had to allow anchorage by the bayonet anchor and winding on the transporting spool (see Figure 5.83).

In the dam, once the old mechanical extensometer was dismounted, the installation of the new optical-fibre extensometer processed rapidly. The extensometer had to be pretensioned *in situ* because of the manner of fixation: the upper anchor piece is fixed to the dam by means of a wedging screw; the lower anchor piece is fixed to the rock using a bayonet anchor, the same as used on the dismounted rod extensometer. Within half a day the sensor was ready to measure. The position of the sensor within the dam is shown in Figure 5.84.

The 30 m long extensometer F3 is placed close and parallel to the 60 m long extensometer F4. In order to compare the results of the measurements, the deformation measured by the extensometer is divided by 2. The optical-fibre extensometer F3 was installed in October 1997.

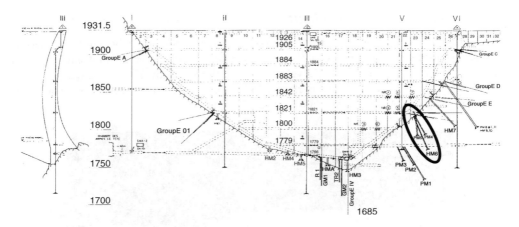

Figure 5.84 Position of optical-fibre and rod extensometers (encircled) in Emosson Dam (courtesy of Emosson).

The F3 measurement data, the measurement data of the rod extensometer F4, the difference between the two extensometers and the stored water level altitude are shown in Figure 5.85. The measurements are performed periodically, once a month, with few exceptions.

The comparison given in Figure 5.85 shows a very good agreement between the two systems. Moreover, the response of the optical-fibre sensor is faster than the rod extensometer response; consequently, the optical fiber sensor is more sensitive than the extensometer. The lag of the rod extensometer with respect to the optical-fibre extensometer was expected, as this is a consequence of its friction on the walls of the borehole in which it is placed. The biggest delay of the extensometer is registered just after the dam level starts to increase. At this time the force of friction changes sign, increasing the duration of inactivity of the rod extensometer.

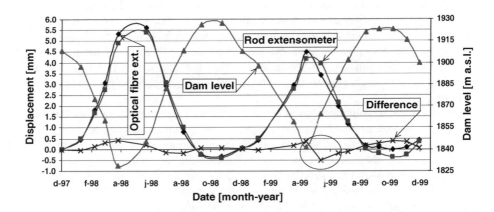

Figure 5.85 Displacement measurements performed with optical-fibre and rod extensometers, as well as their difference and water level in the dam accumulation (Glišić, 2000).

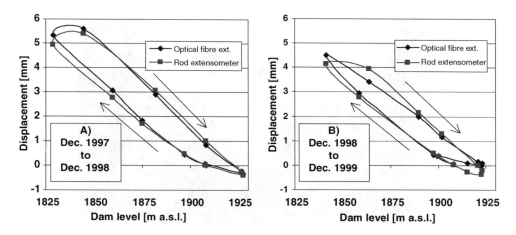

Figure 5.86 Measurements of optical-fibre extensometer F3 and rod extensometer F4 with respect to the water level in dam accumulation (Glišić, 2000).

This effect is even more pronounced in the second year of monitoring (see encircled area in Figure 5.85).

The fact that the optical-fibre extensometer is more sensitive than the rod extensometer is more noticeable in Figure 5.86, which shows the dependence of deformation with respect to the water level. The hysteresis of the optical-fibre extensometer being smaller than that of the rod extensometer indicates a better sensitivity of the optical-fibre extensometer.

A simple application of an optical-fibre interferometric extensometer is presented. The measurements obtained by the optical-fibre extensometer are in agreement with a conventional rod extensometer. They are more accurate, since the rod extensometer has a delay in response to dam movement generated by friction of the rod with the inner borehole walls.

The second example of on-site monitoring of dam parameters is related to average strain and temperature monitoring in the dam during the rising of its crest (Inaudi *et al.*, 1999b). Both discrete and distributed techniques were combined. The Luzzone Dam is located on the small River Brenno di Luzzone, near Olivone, Switzerland. Luzzone is an arch dam with a height of 225 m and crest length of 600 m. It was built in 1963, but its crest was raised by 17 m in 1997–1998 in order to increase its capacity. A view of the Luzzone Dam is given in Figure 5.87.

The volume of accumulation is now 108×10^6 m^3 and the altitude of the crest is at 1609 m a.s.l. In order to upgrade the dam, the new concrete was added to the existing structure. Interaction between the new and existing concrete had to be guaranteed, since the interface zone represented a potential source of problems. Incompatibility of deformation caused by early age deformation generated by temperature development in new concrete and later by the shrinkage of the new concrete could provoke delamination and loss of structural performance.

It was decided to perform monitoring during the construction and in the long term. The deformations of new concrete were monitored using discrete optical-fibre sensors. The sensor network is shown in Figure 5.88.

Figure 5.87   View of Luzzone Dam (courtesy of SMARTEC).

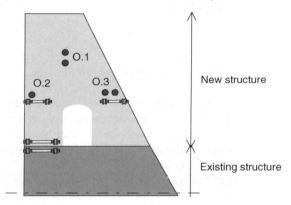

Figure 5.88   Position of deformation sensors in the new and existing structures of Luzzone Dam (courtesy of SMARTEC).

Figure 5.89   Embedding of sensor in the concrete block (courtesy of SMARTEC).

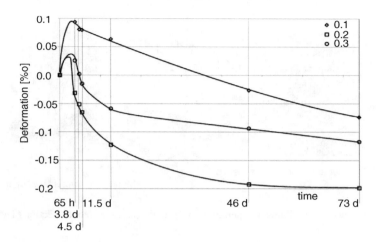

Figure 5.90   Example of results of average strain monitoring (Inaudi *et al.*, 1999b).

Sensors were embedded in the concrete by simply placing at designed locations and covering with a concrete layer. The embedding of the sensor in the concrete block is shown in Figure 5.89.

The results of monitoring are not fully disclosed, but an example registered during 73 days after pouring is given in Figure 5.90. The results show strong dependence of the drying shrinkage on the distance from the concrete surface. The sensors close to the air show a much quicker drying and thermal shrinkage development than those in the centre of the concrete block.

The temperature gradients generated by hydration heat in big concrete blocks of the dam create significant tensional stresses that may cause cracks and compromise the performance of the new structure. In order to evaluate the temperature gradients it was decided to monitor temperature in a large horizontal section of the new concrete using a distributed temperature sensor (Thévenaz *et al.*, 1998).

The 300 m long temperature-sensing cable was installed in a horizontal serpentine spanning the 15 m × 20 m block. The cable was installed in the middle of a 3 m high concrete rise. The selected concrete block was one of the largest in this construction phase and rested directly on the rock at one end of the dam. The position of the distributed temperature sensor in the dam's section is shown in Figure 5.91.

Right after pouring, the temperature is constant in the whole block. After 15 days, a maximum temperature of about 55 °C is reached in the middle of the block, whereas on the boundary the temperature is approximately 25 °C; the maximum temperature difference is 30 °C. This gradient is a result of hydration in the monitored block, but also because of the pouring of an additional concrete block on top of it, 13 days later. The second block brings new heat into the monitored block and also acts as insulation. In the following months, the concrete gradually cools down, and after 6 months the temperature becomes uniform in all the monitored area. The measurements performed at 0, 15 and 183 days after pouring are presented in Figure 5.92.

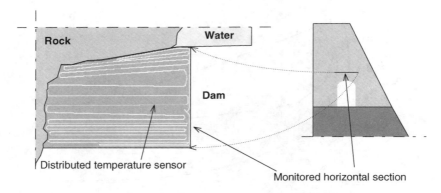

Figure 5.91 Position of distributed temperature sensor in the dam's horizontal section (Thévenaz *et al.*, 1998).

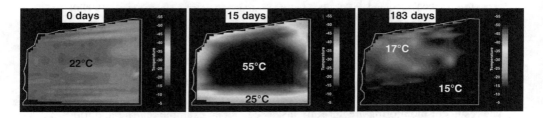

Figure 5.92 Measurements of temperature performed at 0, 15 and 183 days (Thévenaz *et al.*, 1998).

An example of combined average strain and temperature monitoring of the dam using discrete and distributed optical-fibre sensors was presented. Average strain measurements were performed using long-gauge deformation sensors embedded in concrete. The sensors provided for strain distribution in the concrete block due to drying shrinkage.

Distributed temperature sensors provided for approximately 300 temperature values in an area of 15 m × 20 m. The measurements indicated a high temperature gradient 15 days after the pouring of concrete. The temperature gradient practically disappeared only after 6 months. The temperature-sensing cable was installed in less than 1 hour by three persons and caused no delay in the construction schedule.

5.5.4 Monitoring of a Gravity Dam

A gravity dam is commonly built on a large river with a large water flow rate and it may be founded on a soil with relatively bad mechanical properties. It is built of nonreinforced concrete and commonly has a quasi-triangular cross-section, being very wide at the base and relatively narrow at the top. A typical gravity dam cross-section is presented in Figure 5.93.

Gravity dams oppose the water mass by their weight; thus, one of the most important parameters to monitor is the general stability of the dam, namely horizontal displacement and rotation. The gravity dam is conceived to accept only compressive stresses (nonreinforced concrete),

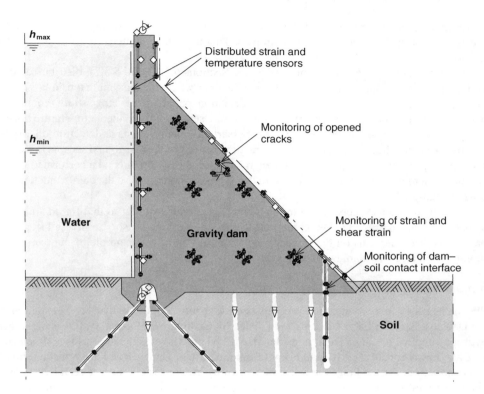

Figure 5.93 Schematic representation of optical-fibre sensor network installed in the cross-section of a gravity dam (courtesy of SMARTEC).

thus it is very important to detect tensions that can generate cracks, for example by water level changes (uplift pressure), by temperature gradients generated during the construction or by environmental temperature variations, and by the alkali reaction in concrete aggregates. Control of seals and drainage is also of importance; therefore, the same parameters as for an arch dam (presented in Section 5.5.2) are to be monitored. The general schema of an optical-fibre sensor network in a gravity dam is shown in Figure 5.93.

Multipoint extensometers provide for monitoring of absolute horizontal and vertical displacement of the dam and strain distribution in the soil. Based on the assumption that the cross-section of the dam moves as a rigid (nondeformable) body, the extensometers can also provide for monitoring of global rotation of the dam. Global rotation is also monitored using inclinometers.

A long-gauge sensor provides for monitoring of the average strain distribution. The evaluation of dam–soil contact stress is of particular interest for the stability of the dam. Since significant shear strain can occur, crossed topologies can also be used in order to determine the principal strain axes and tensions. Opened cracks are monitored using triangular topologies, as presented in Section 4.5.3. The number and position of sensors depends on the project requirements. Temperature sensors are necessary for thermal strain monitoring. For dams situated in seismic areas, it is also important to install accelerometers in inspection galleries.

Monitoring of pore pressure in the soil is also very important, notably in soils with bad mechanical properties (sand, clay and similar). Therefore, piezometers are to be installed in the soil.

In order to perform integrity monitoring (see Sections 5.4.13 and 5.5.2), distributed strain and temperature sensors can be installed along the lines where the maximum tensional stresses are expected. These sensors provide for crack detection, but also for average strain monitoring, and can replace discrete deformation and temperature sensors. Distributed temperature sensors can provide for seepage detection, too. If necessary, a separate seepage detection distributed sensor can be installed along the seals. Schematic positions of the distributed sensors in a cross-section are presented in Figure 5.93. In general, integrity monitoring can be performed on any structural element of the dam whose failure can seriously compromise the dam's function (see on-site example below).

The view of a sensor network from the downstream side is similar to that for an arch dam, as presented in Figure 5.82. The number and position of all the sensors to be used in gravity dam monitoring depends on the project requirements. An on-site example of the parameters monitored in a gravity dam is given next.

5.5.5 On-Site Example of Monitoring a Gravity Dam

Plaviņu hes is a dam belonging to the complex of three most important hydropower stations on the Daugava River in Latvia (Inaudi and Glišić, 2006). In terms of capacity, this is the largest hydropower plant in Latvia and is considered to be the third level of the Daugavas hydroelectric cascade. It was constructed 107 km distant from the Firth of Daugava and is unique in terms of its construction: for the first time in the history of hydro-construction practice, a hydropower plant was built on clay–sand and sand–clay foundations with a maximum pressure limit of 40 m of water. The hydropower plant building is merged with a water spillway. There are 10 hydro-aggregates installed at the hydropower plant and its current capacity is 870 000 kW. The entire building complex is extremely compact. A view of the Plaviņu hes Dam is given in Figure 5.94.

Figure 5.94   View of Plaviņu hes Dam (courtesy of Latvenenrgo).

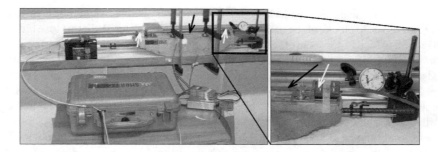

Figure 5.95 View of mechanical testing setup; black arrows point to distributed sensor and white arrows to anchoring pieces (courtesy of SMARTEC).

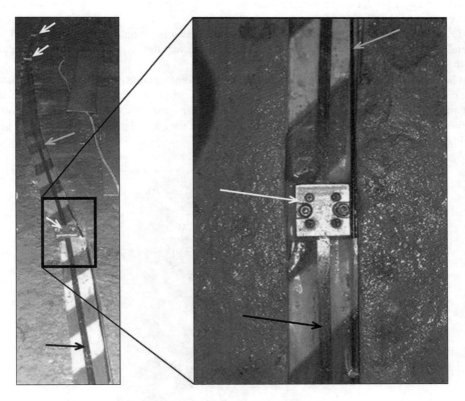

Figure 5.96 View of deformation sensor (black arrow), temperature sensor (grey arrow) and anchor piece (white arrows) installed in the 'trench' (courtesy of Latvenenrgo and VND2).

One of the dam inspection galleries coincides with a system of three bitumen joints that connect two separate blocks of the dam. Owing to abrasion by water, the joints lose bitumen and redistribution of loads appears in the concrete arms. Since the structure is nearly 40 years old, the structural condition of the concrete can be compromised due to ageing. Thus, the redistribution of loads can provoke damage of the concrete arm and, as a consequence, inundation of the gallery. In order to increase the safety and enhance the management activities, it was decided to monitor the integrity of the joint and the average strain in the concrete arm next to the joints. A distributed sensing system was selected for this purpose.

On-site conditions, and notably high relative humidity, imposed specific installation procedures for the distributed deformation sensor. The sensor could not be glued over all the length, but it was clamped to the concrete every 3 m. For this purpose, special anchoring pieces were developed and, before the implementation on site, the installation procedure was tested in the laboratory. The mechanical testing setup is presented in Figure 5.95.

The expected yearly temperature variations in the inspection gallery are approximately 10 °C, and a distributed temperature-sensing cable was installed in parallel with the distributed deformation sensor in order to perform deformation sensor compensation and distinguish thermal and elastic strain components.

The distributed sensors were installed in a purpose built 'trench', which was closed with metallic plates after the installation with the aim to protect the sensors. The sensors installed in the trench are shown in Figure 5.96.

The sensors are read from the central measurement point situated in the dam. Management of data and generation of prewarnings and warnings are performed by the software continuously running on the reading unit. The prewarning and warning messages are sent via a local Ethernet network and a single pole, single throw line. A view of the central measurement point is given in Figure 5.97.

Figure 5.97   View of central measurement point (courtesy of Latvenenrgo).

An example of bitumen joint integrity monitoring in an inspection gallery of a gravity dam was presented in this section. The system general performance and installation issues were also presented. The results of monitoring are not disclosed and, therefore, are not presented, but no indication of damage was detected sofar.

5.5.6 Monitoring of a Dyke (Earth or Rockfill Dam)

A dyke is a large barrier built in order to retain large water accumulations of large rivers, lakes or sea seaboards. There are different types of dyke, depending on their construction manner. In general, the main components are a watertight clay core, a filling material (such as earth or rocks) and, close to the surface, it may have a filtering layer and a reinforced concrete jacket.

Dykes are frequently founded on a soil with relatively bad mechanical properties. They have a trapezoidal cross-section, being very wide at the base and relatively narrow at the top. The angles of the slopes depend on the construction material and are imposed by stability conditions. A typical dyke cross-section is shown schematically in Figure 5.98.

Being a barrier for large water accumulations, dykes must be monitored. The main objectives of monitoring are early detection of slope instability, uncontrolled seepage and piping or internal erosion due to seepage. Uncontrolled seepage can be a consequence of cracking in the concrete jacket or clay core generated by water pressure combined with long-term settlement of dyke materials. That is why the deformation in the jacket, core and soil must be monitored.

The stability of slopes depends mainly on the construction material, the water level table in the dyke itself and the pore pressure in the soil. Pore pressure, therefore, is an important parameter to be monitored.

Multipoint extensometers provide for absolute horizontal and vertical displacement of the dam base and strain and settlement distribution in the soil and clay core. In addition, they provide for crack detection and localization in the clay core.

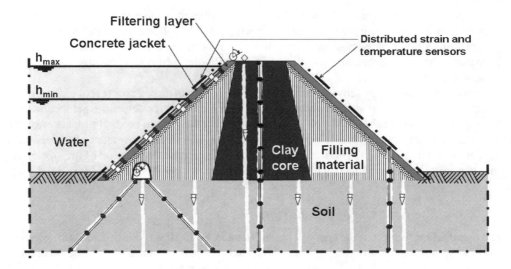

Figure 5.98 Schematic representation of optical-fibre sensor network installed in the cross-section of a dyke (courtesy of SMARTEC).

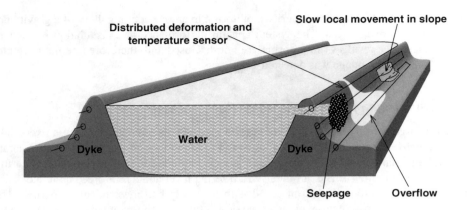

Figure 5.99   Schematic representation of distributed deformation and temperature sensors positions in a dyke for seepage and overflow detection (temperature sensor) and localization of slow slope movements (deformation sensor) (courtesy of SMARTEC).

Long-gauge deformation sensors provide for the average strain and curvature distribution in the concrete jacket, evaluation of the deformed shape and crack detection and characterization. Temperature sensors are necessary to determine thermal gradients and for thermal strain monitoring. Inclinometers provide for absolute rotation in the gallery and on the top of the crest. For dams situated in seismic areas it is also important to install accelerometers in inspection galleries. In order to monitor pore pressure, a number of piezometers are installed along the base of the cross-section. The total number and the positions of sensors depend on the project requirements.

A schematic representation of a sensor network installed in the cross-section of dyke is given in Figure 5.98.

Besides the discrete sensors presented above, a distributed sensing system can be installed in order to perform integrity monitoring (see Sections 5.4.13 and 5.5.2). This system provides for crack detection and average strain monitoring, and can replace discrete deformation and temperature sensors. Schematic positions of the distributed sensors in the cross-section are presented in Figure 5.98.

Distributed temperature sensors installed along the dyke can provide for seepage or overflow detection, whereas distributed deformation sensors can provide for detection and localization of slow movements in the slopes. Seepage or overflow changes the thermal properties of the soil, which are detected by the temperature sensor; the slow local movement of a slope puts the deformation sensor in tension. In both cases, localization and evaluation of the size of the area involved can be performed. The principles are shown schematically in Figure 5.99.

5.5.7   On-Site Example of Monitoring a Dyke

The Canales Dam is built on the Genil River, in Granada, Spain. It was constructed during period 1975–1989, and exploitation started in 1989. The dam is 156 m tall and the crown is 340 m long. The dam accumulation is used to produce electrical power, but it is also an important tourist attraction, notably for people who like fishing. The Canales Dam controls the river flow during

Figure 5.100   View of Canales Dam (courtesy of IIC - Ingenieria de instrumentacion y control).

the year, ensuring that there is always sufficient water to keep the river 'alive' even during the driest months of July and August. A view of the Canales Dam is given in Figure 5.100.

The aim of the monitoring has been crack detection and deformation of the stiff-clay core of the dam. The zone where the cracks can occur was estimated to be between 60 and 110 m deep in the clay core. This whole length was equipped with a 50 m long multipoint extensometer with 10 measurement zones (10 enchained 5 m long deformation sensors). Preparation of the extensometer for the installation is shown in Figure 5.101.

Figure 5.101   A 50 m long multipoint extensometer ready for the installation in the borehole (courtesy of IIC - Ingenieria de instrumentacion y control).

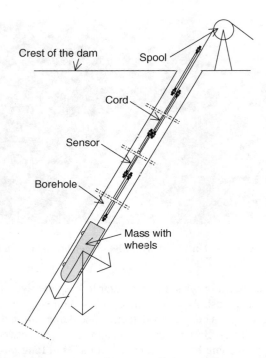

Figure 5.102   Schema of the installation procedure (courtesy of SMARTEC).

The sensors were installed in a 110 m deep and 30° inclined borehole using a purpose-developed installation procedure and special setup. A mass with wheels was attached to the low-stiffness cord and slipped into the borehole. While the mass was pulling the cord into the borehole, the extensometer was attached to the cord, so the cord was used as a guide for the extensometer. The installation procedure is shown schematically in Figure 5.102.

After the placement of the extensometer in the desired position, the borehole was alternatively filled with grout on the anchor pieces and sand over the active zone, in order not to perturb the strain field in the clay core and to guarantee good deformation transfer from the clay core to each deformation sensor making a part of the extensometer. The results of the monitoring are not disclosed.

An application of a multipoint optical-fibre extensometer for measurement of deformation distribution and crack detection in the stiff-clay core or the earth dam (dyke) has been presented. The extensometer, consisting of 10 deformation sensors, each 5 m long, was successfully installed in a 30° inclined borehole at a depth between 60 and 110 m and properly functioning under a pressure of 10 bar.

5.6 Monitoring of Tunnels

5.6.1 Introduction

Being underground objects, tunnels are monitored during their whole lifespan. The selection of monitoring instrumentation, as well as the pace (frequency) of monitoring, depends on the stage of the tunnel, the tunnel type, the soil mechanical quality, the tunnelling method during construction and the purpose of the tunnel.

Concise, but not exhaustive, classifications of tunnels can be made: by stage (existing or new); by cross-section shape (circular, horse-shoe or rectangular); by type of construction (cast on site or prefabricated); by tunnelling method (bored by drilling and blasting, drilled by roadheader excavation, drilled using a tunnel boring machine (with placing of prefabricated tunnel segments or with shotcrete, using the new Austrian tunnelling method), immersed tube tunnelling or cut-and-cover tunnelling); by purpose (pedestrian or cycle tunnels, rail and road tunnels, water tunnels, service tunnels (e.g. telecommunications cabling) and some special tunnels (e.g. particle accelerators)).

The parameters that, in the most representative manner, reflect the tunnel structural behaviour are convergence, strain and deformation in tunnel walls and soil, cracks and joints openings, pore pressure in the soil and the contact pressure (load) between the tunnel and soil or rock.

As for dams, tunnel monitoring strategies are well developed and based mainly on conventional measurement instruments. Monitoring strategies based on the use of discrete optical-fibre sensors are similar to conventional ones: the discrete traditional sensors can be exchanged with optical-fibre sensors that bring advantages in terms of insensitivity to harsh environmental conditions (corrosion, humidity, electromagnetic fields, . . .), accuracy, stability and long-term performance.

The innovation is the use of distributed sensors for average strain, temperature and integrity monitoring. A distributed sensing system is, by its nature, very suitable and efficient for the monitoring of tunnels, these being structures with lengths that can achieve several kilometres.

The monitoring strategies for the most representative types of tunnel and most frequently required monitoring parameters are presented below. The monitoring strategies presented are focused on parameters that can be monitored with optical-fibre technologies. That is the reason why the on-site examples presented below are related to monitoring single parameters, rather than to the monitoring of tunnels as a whole.

5.6.2 Monitoring of Convergence

The most indicative parameter to monitor in the case of tunnels is convergence of the cross-section. Depending on the cross-section shape and the stage of the tunnel at the moment of installation of the monitoring system (new structure: embedding of sensors is possible; existing structure: only surface mounting is possible), different sensor networks can be adopted.

The convergence can be monitored using multipoint extensometers installed perpendicular to the elastic line of the tunnel's cross-section, as shown schematically in Figure 5.103.

This method can be applied to both new and existing tunnels. Besides convergence monitoring, this method provides for average strain distribution in the surrounding soil. Although apparently simple and undemanding, this method may be relatively expensive, since it requires the drilling of the boreholes.

For new tunnels, cast on site or built of prefabricated segments, regardless of the cross-sectional shape, it is possible to use a parallel topology installed in the cross-section in order to evaluate convergence. The parallel sensors provide for curvature and axial strain distribution along the elastic line of the tunnel's cross-section. Temperature sensors are necessary if the

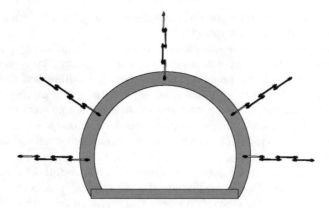

Figure 5.103 Schematic representation of convergence monitoring using multipoint extensometers, applicable to both new and existing tunnels (courtesy of SMARTEC).

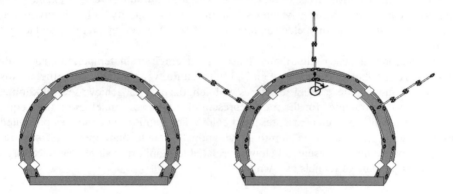

Figure 5.104 Schematic representation of sensor networks for convergence monitoring of new tunnels using only parallel topologies (left) and using combined parallel topologies and multipoint extensometers (right) (courtesy of SMARTEC).

temperature variations and gradients are expected. The deformed shape of the tunnel is evaluated using the double-integration algorithm of curvature and axial deformations, similar to that used for monitoring arches. The deformed shape of the tunnel practically provides for the convergence measurement. The sensor network of parallel topologies is shown schematically in the left-hand side of Figure 5.104.

The tunnel is practically a shell structure; therefore, the longitudinal deformation (along the axis of the tunnel) and curvature influence the evaluation of the deformed shape. Since they are not taken into account, then, unless measured, a certain error is introduced in the calculations. The error can be corrected if parallel topologies are also used in the longitudinal direction, using relatively complex algorithms.

Although, the use of parallel topologies installed only along the elastic line of the tunnel's cross-section provide for results that may be of limited accuracy in terms of traditionally measured convergence, it has a number of advantages:

1. It provides for average strain and average curvature monitoring
2. It provides for crack detection and characterization
3. Besides the convergence, it also provides for relative tangential displacements
4. Sensors are embedded – well protected and invisible
5. No need for boreholes, which decreases the costs
6. Permanent installation.

In order to increase accuracy and make possible the evaluation of the absolute displacements, parallel topologies can be combined with multipoint extensometers and an inclinometer. Since the parallel topologies provide for evaluation of the deformed shape, the number of multipoint extensometers can be limited. The combined sensor network is given in Figure 5.104 (right).

Convergence monitoring of existing tunnels using optical-fibre sensors can be performed using the strategy shown in Figure 5.103. This strategy can be complemented with sensors installed on the tunnel's walls parallel to the elastic line (tangential to the radius of the tunnel), as shown in Figure 5.105 (left). The advantage of this solution is the possibility of monitoring relative tangential movements.

Parallel topologies cannot be applied on existing tunnels since the tunnel–soil interface is not accessible. However, a result similar to that obtained using parallel topologies can be achieved using a combination of deformation sensors and inclinometers installed on the surface of the tunnel walls. The sensors provide for monitoring of average strain on the wall's surface, while inclinometers provide for monitoring rotations of cross-sections at the extremities of the sensors. Knowing the rotations at the extremities of the sensor and the deformation of the wall surface, it is possible to determine the deformation at the tunnel–soil interface. This strategy is presented schematically in the right-hand side of Figure 5.105.

In this section, strategies for monitoring the convergence of tunnels with a horse-shoe cross-section are presented, but similar strategies are applicable for tunnels with other types of cross-section. A simple on-site example of convergence monitoring is presented next.

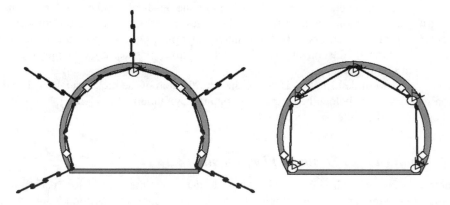

Figure 5.105 Schematic representation of sensor networks for convergence monitoring of existing tunnels using multipoint extensometers combined with surface sensors (left) and using inclinometers combined with surface sensors (right) (courtesy of SMARTEC).

5.6.3 On-Site Example of Monitoring of Convergence

A new inspection gallery of the Luzzone Dam (see Section 5.5.3) was excavated by blasting (Casanova, 1998). The work consisted of four phases, as shown in Figure 5.106 (Inaudi *et al.*, 1998b). Monitoring of convergence was performed during the excavation using the strategy presented in Section 5.6.2 and shown schematically in Figure 5.103. The thickness of shotcrete was determined from the convergence measurements.

After excavation of zone 1, a two-point extensometer was installed horizontally and vertically. After zone 2 was excavated, very small deformations on the two installed extensometers were measured. Two additional two-point extensometers were then installed horizontally and vertically in zone 2. The sensor network is shown in Figure 5.106.

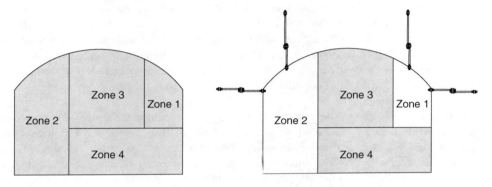

Figure 5.106   Excavation sequences in the Luzzone inspection tunnel (left) and state of the works at the moment of installation of extensometers (right) (courtesy of SMARTEC).

The installation of extensometers was relatively simple: they were attached to grouting pipes and put into the borehole. Then they were fixed to the rock by grout. The excavation work and installation of sensors are shown in Figure 5.107.

The convergence generated after the blasting of zone 3 was of particular interest, since the vault of the cross-section was then fully loaded. Blasting occurred less than 1 m away from the shotcrete-protected sensors ends. Immediately after blasting it was possible to measure the sensors and to evaluate the convergence. The deformation measured after the blasting of zone 3 is shown in Figure 5.108.

The asymmetry of deformation is due to the stratification of the rock formation. The deformation measured was below the estimation obtained by numerical modelling. Therefore, the thickness of the shotcrete was decreased.

5.6.4 Monitoring of Strain and Deformation

Tunnels are shells that curve in two directions. To determine the state of strain at a point it is necessary to use three parallel topologies, as in the case of slabs (see Section 5.3.4). As in the case of slabs, fully equipping the tunnel with parallel topologies will require a large number of sensors and involve significant cost. That is why only the most critical cross-sections are commonly monitored. Where the shear strain is not expected to be significant, the number of

Figure 5.107   Luzzone inspection gallery excavation works (left) and installation of optical-fibre extensometers (right) (courtesy of SMARTEC).

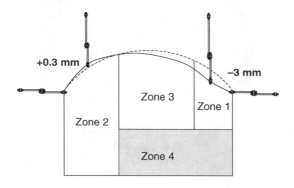

Figure 5.108   Convergence measured after the excavation of zone 3 (courtesy of SMARTEC).

parallel topologies in a single cell can be reduced to two; that is, the inclined parallel sensors can be omitted (see Figure 5.14). An example of a sensor network for tunnel deformation monitoring with a reduced number of parallel topologies in cells is presented in Figure 5.109.

Depending on the project requirements, the cross-sections can be monitored in only one direction, using parallel topologies in new tunnels, as shown in Figure 5.104 (left) or, in the case of existing tunnels, using a combination of surface sensors and inclinometers, as shown in Figure 5.105 (right). Deformation monitoring can be performed independently of convergence monitoring and concentrated only on some particular points in the cross-section. For example, for tunnels with a cross-section built of prefabricated segments that are assembled on site by post-tensioning it may be of interest to monitor deformation around the joints and in the

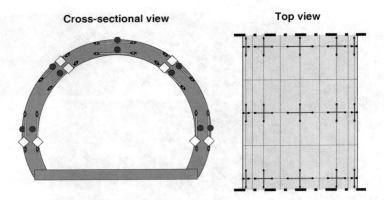

Figure 5.109   Example of a sensor network for tunnel deformation monitoring (without shear strain monitoring) (courtesy of SMARTEC).

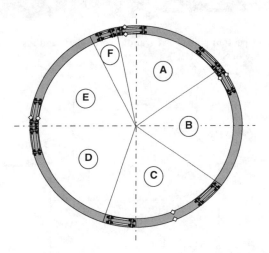

Figure 5.110   Monitoring of deformation of prefabricated cross-sectional segments around the post-tensioned joints and at points where the biggest stresses are expected (courtesy of SMARTEC).

points where the maximum stresses are expected. An example is shown schematically in Figure 5.110.

An alternative solution for existing tunnels is to install parallel topologies on the surface of the structure using specially designed brackets and distancers, as shown in Figure 5.111.

This solution, however, has some limitations: sensor installation and protection require complex equipment and reduce the free opening of the tunnel.

The examples above demonstrate the possibilities of global deformation monitoring of a tunnel's structural elements. In some cases it is of interest to monitor just the average strain in some locations of the cross-section. In those cases, simple topologies are used. An on-site example is presented next.

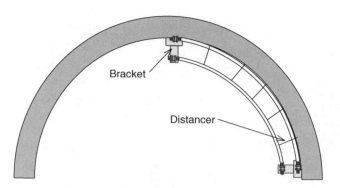

Figure 5.111   Schematic representation of installation of parallel topology on the surface of an existing tunnel (courtesy of SMARTEC).

5.6.5 *On-Site Example of Monitoring of Deformation*

Champ Baly is a cut-and-cover tunnel, built on the A1 motorway, between the Swiss cities of Lausanne and Bern. Construction of the tunnel started in July 1998. The pouring of concrete was completed in August 1999. The tunnel's cross-section consists of two reinforced concrete vaults. The tunnel under construction is shown in Figure 5.112.

The tunnel is 230 m long. Approximately 6000 m^3 of the concrete poured in 21 stages was necessary to construct the tunnel. An ordinary concrete was implemented in the first nine stages and a high-performance concrete in the last 12 stages.

The Champ Baly cut-and-cover tunnel was a pilot project for the application of high-performance concrete in Switzerland. IBAP-EPFL developed a research project to assess the potential durability gains from the use of concrete with silica fume and performed the measurements that are presented in this section.

Figure 5.112   View of Champ Baly cut-and-cover tunnel during construction (courtesy of IBAP-EPFL).

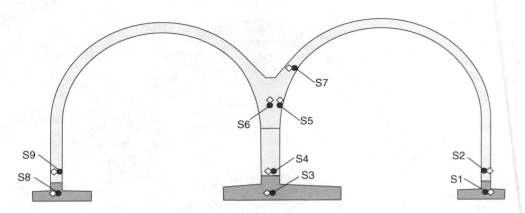

Figure 5.113   Position of sensors in a tunnel cross-section (Glišić, 2000).

Long-gauge deformation optical-fibre sensors were selected for the lifespan monitoring of longitudinal deformations in two reference sections of the tunnel, one built of ordinary concrete and the other built of high-performance concrete. The results obtained on the ordinary section are partially disclosed and presented here.

The section has been equipped with nine 4 m long-gauge optical-fibre sensors and nine thermocouples embedded in concrete, as shown schematically in Figure 5.113. In the figure, the sensors are represented by 'S' followed by a number (S1, S2, etc.).

Sensors S8, S3 and S1 were installed in the foundations poured approximately 3 months before the vaults. The other sensors were embedded in the vaults and walls a few days before the pouring. All the connecting cables were guided to the connection box, which was connected to the reading unit placed 20 m away from the tunnel. During the work, the connection box was installed temporarily. After the works were completed the box was permanently installed in the tunnel.

The first measurements were performed on the foundations, just after the pouring. Sensor S1 was damaged during the pouring. The early age deformation of the foundation was not of interest; therefore, only periodic measurements were done during this period. The average strain measured in the foundations over more than 6 months is presented in Figure 5.114.

Four different periods in the average strain evolution are identified in Figure 5.114. Thermal swelling generated in the very early age was followed by shrinkage. The pouring of the walls and vaults generated additional deformation of the foundations. After pouring, the foundations' deformation is stable and follows the thermal variations of the environment.

Early age deformation in the vaults and walls was monitored for 7 days after pouring of the concrete. Measurements were recorded automatically every 30 min. The results are presented in Figure 5.115. The measurements were then performed with 10-day intervals. An example is given in Figure 5.114.

Three different periods are identified in Figure 5.115: a pouring period, a thermal swelling period and a shrinkage period. During pouring, relatively small deformation due to the new concrete dead load is registered by sensors S8 and S3, installed in the foundations. The average strain measured in the foundations is significantly smaller than that measured in the lower part of the walls by sensors S9, S4 and S2. This is a consequence of the plasticity and rheologic

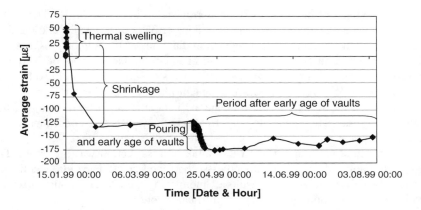

Figure 5.114   Typical average strain evolution in foundations (Glišić, 2000).

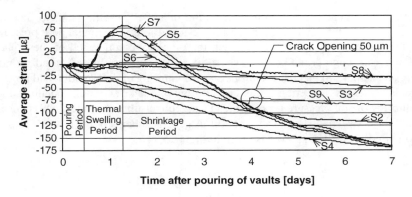

Figure 5.115   Average strain evolution in a tunnel cross-section during the early age of vaults and walls (Glišić, 2000).

properties of fresh concrete. The average strain higher in the vaults, measured by sensors S6, S5 and S7, was very small due to the reduced load at these locations. The setting time was retarded and the new concrete was in its 'dormant' phase during the pouring period.

During the thermal swelling period, the hydration process was activated. The deformation is generated by the temperature development. The average strain in the foundations was small, it was higher in the lower parts of walls and finally, the average strain was very important in the vaults. The average strain in the lower parts of the walls was restrained by the foundations and that is the reason they were smaller than in the vaults. The vaults were not confined and the concrete in these zones could swell significantly more. During this period the concrete is set and stresses are generated in the vaults.

During the shrinkage period the average strain generally decreased, mostly due to thermal shrinkage of the concrete. During this period, a 50 μm wide crack was detected by sensor S9 and visually confirmed on the structure. The crack is a consequence of the vault–foundation interaction and the early age deformation generated by thermal stress.

Different behaviours of the tunnel's parts are registered in Figure 5.115. The first type is related to the foundations, the second type to the lower parts of the walls, and the third type to the vaults. After the cooling, the average strain diagrams of the foundations and the lower parts of the walls are mutually parallel and confirm a monolithic behaviour of the foundations and the walls. The average strain diagrams of the vaults are not parallel to the other diagrams and indicate that the cross-section does not remain plane during the early age period. The interaction between the walls and the foundations is confirmed to be monolithic; thus, the distortion of the cross-section generated the stresses in the vaults. As a consequence, cracking of concrete can appear.

Deformation and strain monitoring of the tunnel's cross-section was presented. The very early age deformation, distortion of the cross-section, and crack detection, localization and quantification were monitored. The results of monitoring were later compared with results obtained in the cross-section built of high-performance concrete.

5.6.6 Monitoring of Other Parameters and Tunnel Integrity Monitoring

Besides the convergence and deformation of the tunnels, it is important to monitor other parameters, such as deformation in the soil, cracks and joints openings, pore pressure in the soil and the contact pressure (load) between the tunnel and soil. Each of these parameters is monitored with appropriate sensors whose locations are determined based on numerical modelling and actual soil mechanical properties. Monitoring of deformation and pore pressure in the soil is presented in Section 5.5 and monitoring of cracks and joints openings in Section 4.5. Monitoring of soil pressure is performed using optical-fibre pressure (load) cells installed at the interface between the tunnel and the soil. A general schema of the tunnel's cross-section equipped with optical-fibre sensors is given in Figure 5.116.

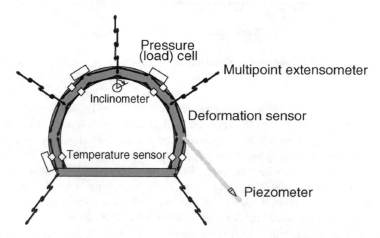

Figure 5.116   Schematic representation of optical-fibre sensors installed in a tunnel cross-section.

Integrity monitoring of bridges is presented in detail in Section 5.4.13, and a similar concept can be applied in the case of tunnels. A schematic example of tunnel integrity monitoring is given in Figure 5.117.

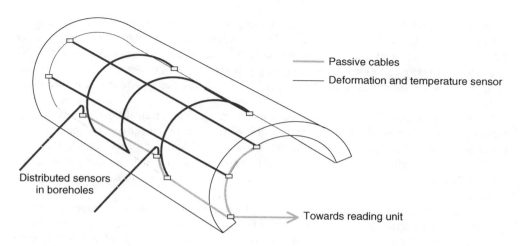

Passive cables

Deformation and temperature sensor

Distributed sensors in boreholes

Towards reading unit

Figure 5.117   Schematic example of tunnel integrity monitoring (courtesy of SMARTEC).

Distributed sensors are installed on the walls and vaults of the tunnel in longitudinal and tangential directions. These sensors provide for crack detection and localization and for detection of local average strain changes due to damage.

In addition, the distributed sensors provide for average strain distributions and temperature monitoring. With adequate installation, they can also replace the multipoint extensometers in the soil (see Figure 5.117) and can be even used for fire detection. Taking into account the number of monitoring parameters and tunnel dimensions, distributed sensing techniques offer convenient, cost-effective and multipurpose instrumentation.

5.7 Monitoring of Heritage Structures

5.7.1 Introduction

Heritage structures are historical constructions that have an important role in the cultural heritage of the society. They may be appreciated for several reasons: age, architectural beauty, religious reasons, historical events or persons they hosted, construction challenges they had in the time that they were built and so on.

Being built many years ago, using old techniques and exposed to environmental conditions for a long time, heritage structures can show different degrees of decay. Without appropriate management, the structure may partially or entirely collapse, which would not only be of concern for safety and economic issues, but will also create an irreversible cultural loss.

The management and the security of heritage structures, in particular, require periodic monitoring, maintenance and conservation. Furthermore, accurate knowledge of the behaviour of a structure is becoming more important as new building and conservation techniques are introduced and the existing constructions are often required to survive undamaged to nearby constructions.

Lack of technical details (drawings, static system) and structural condition makes monitoring of heritage structures very challenging; it is necessary first to 'understand' the structure and its behaviour and then to define a full monitoring strategy. Long-term monitoring helps to increase knowledge of the real behaviour of the structure and in the planning of maintenance intervention. In the long term, static monitoring requires accurate and very stable systems, able to relate measurements often spaced over long periods of time. Furthermore, sensors should be installed with minimal invasion to the aesthetics and functionality of the structure under test.

From many points of view, fibre-optic sensors are the ideal transducers for civil structural monitoring. Being durable, stable and insensitive to external influences, they are particularly interesting for the long-term health assessment of heritage structures. On the other hand, heritage structures are commonly built as masonry, having material and physical discontinuities. Consequently, their structural behaviour is to be monitored with long-gauge sensors.

A general monitoring strategy for heritage structures does not exist because a heritage structure can be a structure of any type: a building (church, palace, etc.), bridge, monument, tunnel and so on. Thus, the strategy developed for the corresponding type of structure is to be used (as presented in previous sections of this chapter), taking into account the particularities of the heritage structures. That is the reason why no particular strategies are presented in this section. However, brief application examples are presented with the aim of highlighting the potential and demonstrating the applicability of optical-fibre sensors in the domain of monitoring of heritage structures. One example of monitoring of a heritage tunnel is presented in Section 4.5.4.

5.7.2 Monitoring of San Vigilio Church, Gandria, Switzerland

San Vigilio church is situated in the small village of Gandria, on Lake Lugano in Switzerland. It was built in the sixteenth–seventeenth centuries, and the frontal façade and portal dates from the nineteenth century. It consists of three boards with niches and a chapel. The post-Romanic bell tower from 1525 attracts by its beauty. Repair and reconstruction work was performed by the Rabaglio brothers in 1771–1786. The interior of the church is covered with frescos.

The church has a significant crack running along the centre of the cylindrical vault. Other smaller cracks are partially present on the upper convex side of the vault (Inaudi *et al.*, 2001). The crack positions are shown schematically in Figure 5.118.

Since the static system of the church and the evolution of the crack openings are not very well known, it was decided to install 10 sensors to monitor the cracks and the curvature variations of the vault. The sensors were 30–50 cm long and attached to both sides of the vault using L-brackets. All optical connections are joined in the sacristy and the sensors are measured

Figure 5.118   Position of cracks and sensors in the vault of the San Vigilio church (cross-sectional view) (courtesy of SMARTEC).

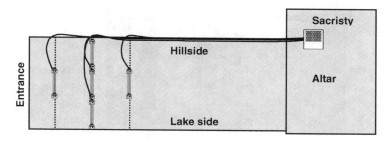

Figure 5.119   Position of the sensors, top view (courtesy of SMARTEC).

Figure 5.120   View of sensors installed in the vaults (courtesy of SMARTEC).

regularly without the need to install scaffolding in the church, which is still in use. The sensor positions are shown in Figures 5.118 and 5.119.

Sensor installation was carried out in just 1 day and the sensors are barely visible to the uninformed observer. Views of the sensor installation are shown in Figure 5.120.

Owing to budget restrictions, and since the deformations were expected to evolve smoothly, it was decided against installing a reading unit permanently in the church. Instead, measurement campaigns of at least 1 week were carried out at intervals of about 3 months. This monitoring programme allows observation of the deformations due to daily and seasonal temperature variations. An example of the results obtained by two of the sensors is shown in Figure 5.121.

The results shown in the figure confirm closing of the observed crack. This indicates a slow redistribution of the loads inside the vault that is the subject of further investigation.

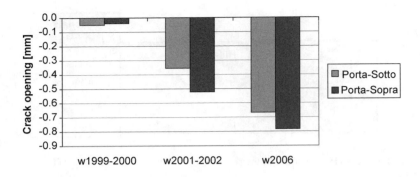

Figure 5.121   Example of results collected on San Vigilio church (data provided by SMARTEC).

5.7.3 Monitoring of Royal Villa, Monza, Italy

Royal Villa in Monza, close to Milan, northern Italy, was built in 1777–1779 by architect Piermarini for Maria-Therese of Austria (Del Grosso *et al.*, 2004). It was modified by the Italian king Umberto I. After the king was killed there in 1900, the villa was no longer used by the royal family and it was practically abandoned during the last decades of the twentieth century. A view of Royal Villa is shown in Figure 5.122.

The following issues have developed over time since being abandoned:

• A system of cracks along the barrel vaults of the central corridor at various levels in the north and south wings
• Degradation of several wooden structures, and notably in the 18 m long truss in the Belvedere.

Italian government and the authorities of Milan and Monza decided to renew the villa and to transform it into a museum. The repair and conservation work had to be done in both the building and the roof. Owing to a complex static system and uncertainties related to structural

Figure 5.122   View of Royal Villa in Monza (courtesy of SMARTEC).

Figure 5.123   Long-gauge sensors installed between the walls in main corridors (courtesy of Tecniter).

behaviour, it was decided to monitor the villa before, during and after the work. Monitoring data were used practically to 'govern' the renewal work (Del Grosso *et al.*, 2004). Both conventional and optical-fibre sensors were used.

Optical-fibre sensors were mainly used as extensometers installed between the walls, orthogonal to the corridor axes, but shorter sensors were also used for crack monitoring. The sensors installed between the walls are shown in Figure 5.123.

Sophisticated modelling and analysis of the data were performed at the University of Genoa. As an example of the results, analysis of the displacement–temperature correlation is presented in Figure 5.124 (Del Grosso *et al.*, 2004). The regression coefficient between displacements and temperature changed in July 2003. High correlation coefficients between displacements and temperature were calculated in both cases. Therefore, it was concluded that the change in regression coefficient indicated that some nonlinear effect occurred.

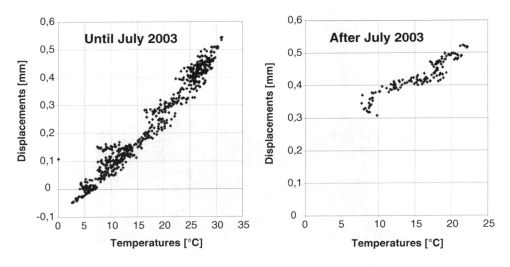

Figure 5.124   Example of results obtained during Royal Villa monitoring (Del Grosso *et al.*, 2004).

The data interpretation and analysis of the wings were carried out statistically, due to the complexity of the structure and uncertainty related to its static system and structural behaviour. The Belvedere main truss displacements were interpreted by means of a numerical model optimized by fitting the observed response with the computed rotations. The monitoring systems installed in the Royal Villa at Monza have allowed a monitoring-based rehabilitation process.

5.7.4 Monitoring of Bolshoi Moskvoretskiy Bridge, Moscow, Russia

Bolshoi Moskvoretskiy Bridge was built in 1936–1937, over the Moscow River (Del Grosso and Inaudi, 2004). It is situated in the centre of Moscow, next to the Kremlin, and is one of the main traffic lines of the city to Red Square. The bridge consists of three parallel 100 m long reinforced concrete arches hidden behind stone walls. The cross-section of each arch contains three merged boxes. The superstructure of the bridge is supported by columns. Four traffic lanes cross the bridge in each direction. A view of the bridge is given in Figure 5.125.

Two types of degradation are noticed on the bridge. Settlement in the centre of the arch, which provoked the cracking of the stone walls near the abutments on both sides of the bridge, and chloride diffusion that practically traverses the upper wall of the arch boxes in some sections and penetrates inside the boxes. The condition of the bridge after nearly 70 years of service and its functional and historical importance have led the authorities to decide to monitor the structural behaviour of the bridge continuously.

The aim of monitoring is to increase knowledge concerning the structural behaviour of this very old structure, to increase safety and to reduce maintenance costs. A total of 16 long-gauge sensors are installed in parallel topologies in order to monitor continuously the average strain along the arch, the curvature in both the horizontal and vertical directions and the deformed

Figure 5.125   View of Bolshoi Moskvoretskiy Bridge; the Kremlin is visible behind (courtesy of SMARTEC).

Figure 5.126 View of sensors installed on the Bolshoi Moskvoretskiy Bridge (courtesy of Triada Holding).

shape. In order to distinguish the thermal influence, six temperature sensors are also installed. Views of the surface-mounted sensors before and after the metallic protection was installed are presented in Figure 5.126.

The bridge is permanently monitored and the data are sent remotely to the control room using a telephone line. The results of the monitoring are not disclosed and, therefore, cannot be presented.

5.8 Monitoring of Pipelines

5.8.1 Introduction

Distributed fibre-optic sensing presents unique features that have no match in conventional sensing techniques. The ability to measure temperatures and strain at thousands of points along a single fibre is particularly interesting for the monitoring of elongated structures, such as pipelines, flow lines, oil wells and coiled tubing. Sensing systems based on Brillouin and Raman scattering are used, for example, to detect pipeline leakages, verify pipeline operational parameters and prevent failure of pipelines installed in landslide areas, optimize oil production from wells and detect hot-spots in high-power cables. Recent developments in distributed fibre-sensing technology allow the monitoring of 60 km of pipeline from a single instrument and of up to 300 km with the use of optical amplifiers. New application opportunities have demonstrated that the design and production of sensing cables is a critical element for the success of any distributed sensing instrumentation project. Although some telecommunications cables can be effectively used for sensing ordinary temperatures, monitoring high and low temperatures or distributed strain present unique challenges that require specific cable designs. The following sections report a number of significant field application examples of this technology, including leakage detection on brine and gas pipelines, strain monitoring on gas pipelines and combined strain and temperature monitoring on composite flow lines and composite coiled-tubing pipes.

Flowlines (i.e. pipelines or gas-lines) often cross hazardous environmental areas from the point of view natural exposures, such as landslides and earthquakes, and from the point of view of third-party influences, such as vandalism or obstruction. These hazards can significantly change the original structural functioning of the flowline, leading to damage, leakage and failure with serious economic and ecologic consequences. Furthermore, the operational conditions of the pipeline itself can induce additional wear or even damage.

Structural and functional monitoring can significantly improve pipeline management and safety. On being provided regularly with parameters featuring the structural and functional condition of the flowline, monitoring can help (1) prevent failure, (2) detect a problem and its position in time and (3) undertake maintenance and repair activities in time. Thus, safety is increased, maintenance cost is optimized and economic losses are decreased. Typical structural parameters to be monitored are strain and curvature, whereas the most interesting functional parameters are temperature distribution, leakage and third-party intrusion. Since flowlines are usually tubular structures with kilometric lengths, structural monitoring of the full extent is an issue itself. The use of discrete sensors, short- or long-gauge, is practically impossible, because it requires installation of thousands of sensors and very complex cabling and data acquisition systems, raising the monitoring costs. Therefore, the applicability of discrete sensors is rather limited to some chosen cross-sections or segments of the flowline, but not extended to full-length monitoring. Other current monitoring methods include flow measurements at the beginning and end of the pipeline, offering an indication of the presence of a leak, but little information on its location.

Recent developments of distributed optical-fibre strain and temperature-sensing techniques based on Brillouin scattering offer promise in providing cost-effective tools to allow monitoring over kilometric distances. Thus, using a limited number of very long sensors it is possible to monitor the structural and functional behaviour of flowlines with a high measurand and spatial resolution at a reasonable cost.

The aim of this section is to present a general approach to the use of distributed sensing for the monitoring of pipelines and relevant on-site application examples.

5.8.2 Pipeline Monitoring

The management of pipelines presents unique challenges. Their long length, high value, high risk and often difficult access conditions require continuous monitoring and optimization of maintenance interventions. The main concern for pipeline owners comes from possible leakages that can have a severe impact on the environment and put the pipeline out of service for repair. Leakages can have different causes, including excessive deformation caused by earthquakes, landslides or collisions with ship anchors, corrosion, wear, material flaws or even intentional damage.

Leakages can be detected and localized using distributed fibre-optic temperature sensors. Fluid pipelines generate a hot-spot at the location of the leak, because the flowing fluid generally has higher temperature than the environment because of friction and initial conditions. In contrast, gas pipelines generate a cold-spot due to the gas pressure relaxation and related gas cooling. These localized thermal anomalies can be detected by a distributed measurement with good spatial, time and temperature resolution, such as those offered by distributed Brillouin and Raman sensing systems.

Furthermore, it is often possible to detect damage even before a critical state is reached. Fibre-optic sensor systems are ideally suited for these tasks. By measuring distributed strain it is possible to determine the increased stresses due to external actions (such as landslides and earthquakes) or to internal causes (such as a reduction of cross-section due to corrosion and wear). Finally, distributed temperature and strain monitoring can detect third-party intrusion before any damage is done to the pipeline.

Distributed Brillouin scattering systems can be used for distributed measurements of both strain and temperature over extremely long distances, limiting the number of instruments that are necessary to monitor a long pipeline.

The positioning of the sensing cable or cables around the pipeline is a critical element for the successful monitoring of a pipeline. In general, the following rules can be established:

- For leakage detection of fluids, the ideal position is below the pipeline at a distance of typically half the pipeline diameter. Leakages at any position around the pipe will generate flows that will tend to descend towards the sensing cable. If an oil pipeline is installed below the water table, then the cable must be placed above the pipe, because the oil will tend to rise. Installing additional sensors on the sides of the pipeline will increase redundancy and the speed of detection.
- For gas leakages, the sensing cable should be placed in thermal contact with the steel pipeline, at any position around it. The leakage will cool down the pipeline and this will be detected by the cable. Installing multiple sensing cables around the pipe will increase detection speed and sensitivity.
- For strain and deformation sensing, the strain-sensing cable must be placed in direct contact with the pipeline steel and firmly attached to it in order to correctly transfer strain. In many cases, it is possible to install the sensor on top of the corrosion protection layer. In order to separate longitudinal strain from bending strain, it is recommended to install three sensors at 120° intervals around the pipeline circumference.
- For intrusion detection, the ideal cable position is above a buried pipeline, halfway to the surface. An intruder will reach the cable before the pipeline, either breaking it or causing a temperature disturbance that can be detected and localized.
- Additional distributed sensing cables can be placed in the ground to detect its movements.

Figure 5.127 summarizes the different recommended cable positions.

Pipeline monitoring systems are based on a combination of sensing cables, measurement instruments and data processing software. Different cables are available for temperature sensing (normal and high temperatures), strain sensing and combined strain and temperature sensing. Specialized software packages are available for detecting leakages from gas, fluids and multiphase pipelines, to display and publish the measurement results in a user-friendly interface and to generate warnings when abnormal conditions are detected.

5.8.3 Pipeline Monitoring Application Examples

The following application examples show how different pipeline monitoring tasks can be addressed.

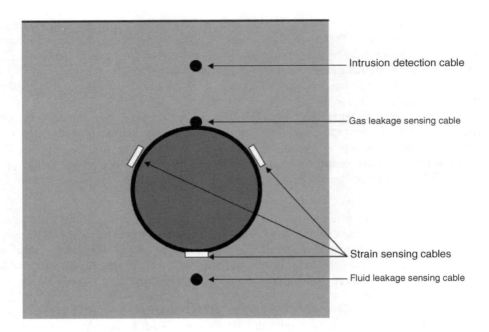

Figure 5.127   Ideal sensor cable placement around the pipe (courtesy of SMARTEC).

5.8.3.1 Leakage Detection in a Brine Pipeline

Using mining technology, the building of underground caverns for gas storage in large rock-salt formations requires hot water and produces large quantities of water saturated with salt, the so-called brine. In most cases the brine cannot be processed on site and must be transported by a pipeline to a location where it can either be used for chemical processes or injected back safely into the ground. Because brine can be harmful for the environment, the pipeline must be monitored by a leakage detection system.

In this example, a 55 km pipeline was built and a leakage detection system using two DiTeSt analysers was installed (Niklès *et al.*, 2004). The installation of the fibre cable required a large number of splices (i.e. 60, which corresponds to an additional loss of up to 3 dB). This reduced the distance range of the instrument accordingly and justified the use of two instruments, since range extender technology was not yet available.

During the construction phase, the temperature sensor was first placed in the trench and buried in the sand some 10 cm underneath the pipeline. The pictures in Figure 5.128 show the construction of the pipeline before and after the pipeline was put in the trench.

The temperature profiles measured by both DiTeSt instruments are transferred every 30 min to a central PC and further processed for leakage detection. Dedicated software performs the leakage detection through a comparison between recorded temperature profiles, looking at abnormal temperature evolutions, and generates an alarm in the case of the detection of leakage. The system is automatically able to transmit alarms, generate reports, periodically reset and restart measurements, and requires virtually no maintenance.

Figure 5.128   Construction phase of a buried brine pipeline. The fibre-optic cable is placed in the sand at the 6 o'clock position about 10 cm underneath the pipeline (Niklès *et al.*, 2004).

The pipeline construction phase was completed in November 2002 and the pipeline was put into operation in January 2003. In July 2003, the first leakage was detected by the monitoring system. It was later found that the leakage was accidentally caused by excavation work in the vicinity of the pipeline. Figure 5.129 shows the occurrence of the leakage and its effect on the

Temperature profile before leakage

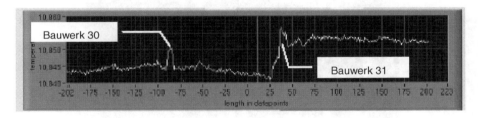

Temperature profile when the leakage is detected

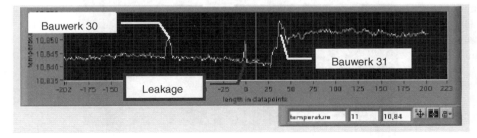

Figure 5.129   Measured profiles before and after leakage (Niklès *et al.*, 2004).

temperature profiles, showing a local temperature increase of 8 °C. An alarm was immediately and automatically triggered and the flow was eventually stopped (Niklès *et al.*, 2004).

5.8.3.2 Gas Pipeline Monitoring

About 500 m of a buried 35-year-old gas pipeline lies in an unstable area. Distributed strain monitoring could be useful in order to improve the vibrating wire strain-gauge monitoring system actually used on the site. Landslides occur in the area over time and could damage the pipeline and put out it of service. Three symmetrically disposed vibrating wires were installed at several sections at a distance typically of 50–100 m, chosen as the most stressed regions according to a preliminary engineering evaluation. These sensors were very helpful, but could not fully cover the length of the pipeline and only provided local measurements.

Different types of distributed sensor were used: SMARTape strain sensors and temperature sensing cable. Three parallel lines constituted of five segments of SMARTape sensor were installed over the whole concerned length of the pipeline as shown in Figure 5.130.

The lengths of segments ranged from 71 to 132 m, and the positions of the sensors with respect to the pipeline axis were at 0°, 120° and −120° approximately. The strain resolution of the SMARTape is 20 $\mu\varepsilon$ with a spatial resolution of 1.5 m (and an acquisition range of 0.25 m) and provides for monitoring of the average strains, the average curvatures and the deformed shape of the pipeline. The temperature-sensing cable was installed onto the upper line (0°) of the pipeline in order to compensate the strain measurements for temperature. The temperature resolution of the sensor is 1 °C with the same spatial resolution and acquisition

Figure 5.130   SMARTape on a gas pipeline (courtesy of SMARTEC).

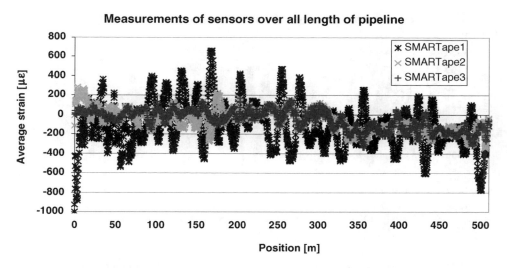

Figure 5.131 Strain distribution over the monitored part of a pipeline measured by SMARTape sensors (courtesy of SMARTEC).

range of the SMARTape. All the sensors are connected to a central measurement point by means of extension optical cables and connection boxes. They are read from this point using a single DiTeSt reading unit. Since the landslide process is slow, the measurement sessions were performed manually once a month. In the case of an earthquake, a measurement session was performed immediately after the event. All the measurements obtained with the DiTeSt system are correlated with the measurements obtained with vibrating wires. At the present stage, the sensors have been measuring for a period of 2 years, providing interesting information on the deformation induced by burying and by landslide progression. A gas leakage simulation was also performed with success using the temperature-sensing cable.

The diagram showing the strain distribution measured by SMARTapes over the whole length of the pipeline after burying is presented in Figure 5.131. The normal cross-sectional strain distribution and the curvature distribution in the horizontal and vertical planes are calculated from the measurements and presented in Figure 5.132.

During the work, the pipe was laid on soil supports every 20–30 m. Therefore, its static system can be considered as a continuous girder. After burying, the pipe was loaded with soil and, therefore, deformed. The pipe cross-sections located on the supports were subject to negative bending (traction at the top part) and the section between the supports to positive bending (traction at the bottom part). The maximum allowed strain in the elastic domain is 1750 $\mu\varepsilon$, and maximum curvature without normal forces 5303 $\mu\varepsilon$ m^{-1}.

During sensor placement and burying of the pipe, an empty plastic tube was installed connecting the upper part of the pipe with the surface, 50 m from the beginning of the first monitored segment. This tube was used to simulate a leakage of gas. Carbon dioxide was injected in to the tube, cooling down the pipe end, due to pressure relaxation, and making the thermal conditions surrounding the contact between the pipe and the tube similar to conditions expected in the case of leakage. This process is presented in Figure 5.133.

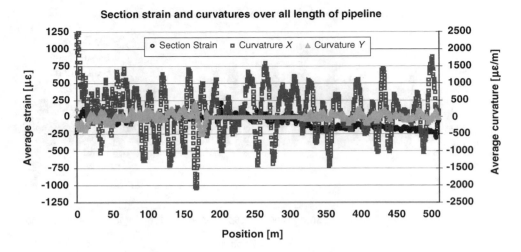

Figure 5.132 Cross-section strain and curvature distribution measured by SMARTape sensors (courtesy of SMARTEC).

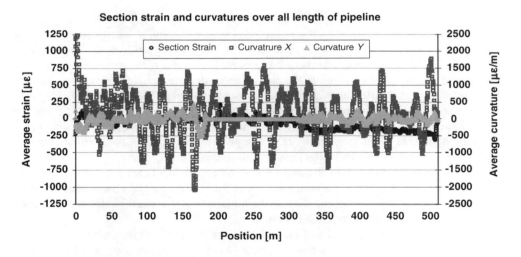

Figure 5.133 Leakage simulation test (courtesy of SMARTEC).

A reference measurement was performed before the leackage. After the carbon dioxide was inserted, the temperature measurements were performed every 2–10 min and compared with the reference measurement. The results of the test are presented in Figure 5.134. The test was successful and the point of simulated leakage is clearly observed in the diagrams (encircled area in Figure 5.134).

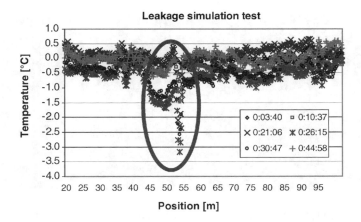

Figure 5.134 Results of leakage test; leakage is detected as a temperature change (courtesy of SMARTEC).

5.8.3.3 Composite High-Pressure Pipe Monitoring: SmartPipe

SmartPipe is a high-strength, lightweight, monitored reinforced thermoplastic pipe that can be used for the rehabilitation of an existing pipeline, or as a stand-alone replacement. The key feature of the technology that underlies SmartPipe is the use of ultra high-strength fibres that are wrapped onto a high-density polyethylene core pipe (see Figure 5.135). Through selection

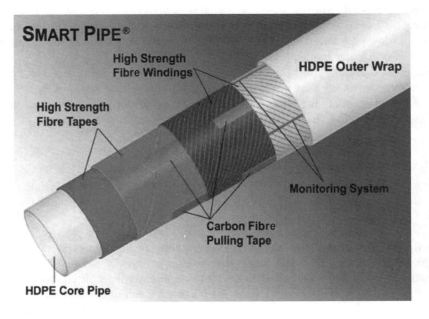

Figure 5.135 SmartPipe design, including SMARTprofile monitoring system (courtesy of Smart Pipe Company).

Figure 5.136   SMARTprofile integration with high-strength fibre windings (courtesy of SMARTEC).

of the fibres, the lay angles, and their sizes, SmartPipe can be specially tailored for any given condition in terms of design pressure, pull-in length (for a rehabilitation), and safe operating duration.

In urban and environmentally sensitive areas it delivers significant savings in terms of costs because it permits access to difficult locations using its trenchless installation methods. It is simultaneously manufactured and installed as a tight-fit liner in up to 50 000 feet of an underground pipeline without any disruption of the surface areas covering the pipeline (except for a small opening at the entry and exit points of the pipeline section being lined). It can restore the subject pipeline to its full pressure service rating, renewing the projected service life of the subject pipeline to an as new or better than new condition, and in most cases does so without reducing the flow rates through the line despite the nominal reduction in inside diameter of the pipeline that occurs due to the presence of the liner.

The integrated SMARTprofile sensors (see Figure 5.136) provide the operator of the pipeline with continuous monitoring and inspection features to assure safe operation of the line throughout the renewed operating life of the pipeline and to provide compliance with the regulations now emerging under the various pipeline safety acts.

5.8.3.4   Composite Coiled Tubing Monitoring

The larger hydrocarbon reservoirs in Europe are rapidly depleting. The remaining marginal fields can only be exploited commercially by the implementation of new 'intelligent' technology, such as electric coiled tubing drilling or intelligent well completions. Steel coiled tubing with an internal electric wire line is the current standard for such operations. Steel coiled tubing suffers from corrosion and fatigue problems, which dramatically restrict the operational life. The horizontal reach of steel coiled tubing is limited due to its heavy weight. The inserted wire line results in major hydraulic power losses and is cumbersome to install. To address these issues, a joint research project supported by the European Commission was started in the year 2000.

The project aims to solve these problems by researching and developing a high-temperature, corrosion- and fatigue-resistant thermoplastic power and data transmission composite coiled tubing (PDT-COIL) for electric drilling applications. This PDT-COIL contains embedded electrical power and fibre-optics for sensing, monitoring and data transmission.

The PDT-COIL consists of a functional liner containing the electrical and the optical conductors and a structural layer of carbon and glass fibres embedded in high-performance thermoplastic polymers. The electric conductors provide electric power for electric submersible pumps or electric drilling motors. A fibre-optic sensing and monitoring system, based on the SMARTprofile design, is also integrated in the liner thickness over its whole length and is used to measure relevant well parameters, monitor the structural integrity of the PDT-COIL and can be used for data transmission (see Figure 5.137).

The embedded optical fibre system was tested for measuring strain, deformations and temperatures of the coil.

Testing of distributed strain and deformation measurements was performed on a 15 m long section of polyethylene liner with integrated strain-sensing fibres. The diameter of the tube was 56 mm. Four optical fibres were installed with angles of $-2.5°$, $-5°$, $5°$ and $10°$ with respect to the tube axis, in order to evaluate the performance of fibres installed with different angles. Two sensors with angles of $-5°$ and $10°$ were connected one after the other and a closed loop was created with the reading unit. The temperature was measured on coils with free optical fibres installed before, between and after the strain-sensing sections.

The aim of this test was to verify the performance of the monitoring system and algorithms. The following tests were performed: traction test, torsion test, combined traction and torsion test, bending test, half tube bending test, double bending test and a combined bending and torsion test.

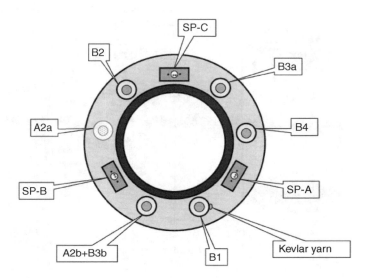

Figure 5.137 PDT-COIL cross-section. The fibre-optic sensing SMARTprofiles are designated by SP-A, SP-B and SP-C (courtesy of SMARTEC).

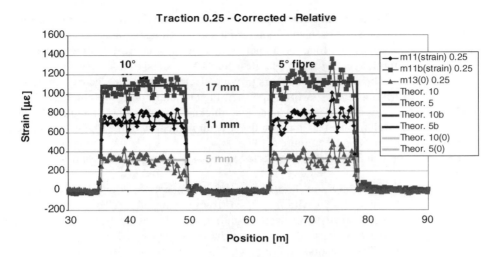

Figure 5.138   Results of the traction test and comparison with theoretical prediction (courtesy of SMARTEC).

The results of this test confirmed the excellent performance of the Brillouin reading unit, providing a resolution compatible with the requirements (better than ±30 με) and short measurement time (better than 5 min). Resolution of temperature was better than 1 °C. As examples, the results of traction and torsion tests are presented in Figures 5.138 and 5.139.

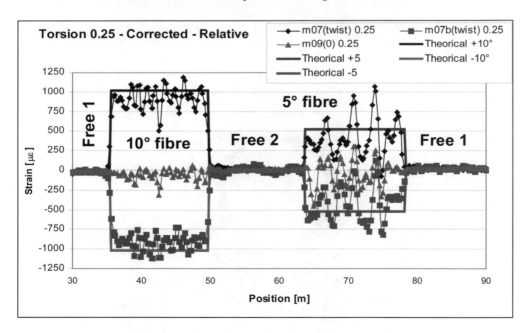

Figure 5.139   Results of the torsion test and comparison with theoretical predictions; higher winding angles provide more sensitivity and accuracy for torsion measurements (courtesy of SMARTEC).

Figure 5.140   Liner heating test by electrical current injection (courtesy of SMARTEC).

To test the temperature-sensing capabilities of the PDT-COIL sensing system, a 150 m section of integrated liner was heated by injecting different levels of current in to the electrical conductors, as shown in Figure 5.140.

Figure 5.141 shows the recorded temperature profile for different current levels. Note that the temperature is not constant along the liner, since one part of the liner was in direct contact with the metallic winding drum that acted as a heat sink, while further sections were wound on a second layer that was essentially surrounded by air and, therefore, thermally insulated. In real applications, the PDT-COIL tubing would be cooled by the fluids circulating inside and outside the pipe.

5.8.4 Conclusions

The use of a distributed fibre-optic monitoring system allows a continuous monitoring and management of pipelines, increasing their safety and allowing the pipeline operator to take informed decisions on the operations and maintenance of the pipe.

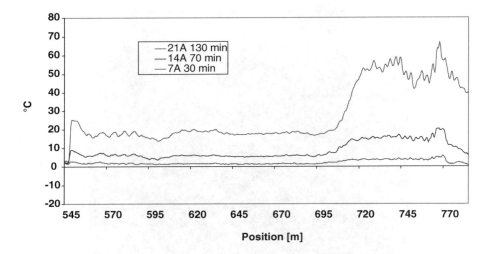

Figure 5.141   Liner temperature changes for different current levels and heating times. The first 545 m
of optical fibre is not integrated in the liner and not shown (courtesy of SMARTEC).

The monitoring system presented and the application examples shown in this chapter demonstrate how it is possible to obtain different types of information on the pipeline state and conditions. In particular, a distributed fibre-optic system allows the following monitoring tasks:

- *Distributed temperature monitoring.* This allows the measurement of the temperature profile
  along the pipe and, therefore, of the temperature changes of the fluid transported. This
  information can be used for optimizing operational parameters and for the identification and
  location of hydrate, ice and wax accumulations. These may be detected by sensing changes
  in temperature on either side of the accumulation.
- *Leakage detection.* Through the identification of temperature anomalies, it is possible to
  detect and localize leakages of small quantities, which cannot be detected by conventional
  volumetric techniques. Furthermore, the ability to pinpoint the exact location of the leak
  allows an immediate reaction at the event location, minimizing downtime and ecological
  consequences.
- *Intrusion detection.* Based on a similar approach, focusing on localized strain and temperature changes, the presence and location of an accidental or intentional intrusion can be
  detected. This enables preventive action before the intruder can damage the pipeline.
- *Distributed strain and deformation monitoring.* This provides information on the strain evolution along the pipeline. This is particularly useful at critical locations, where movements
  caused by earthquakes, landslides, settlements or human activities can introduce potentially
  dangerous strain conditions to the pipeline. Distributed strain monitoring allows the early
  detection of such conditions, allowing an intervention before real damage is produced. This
  is a useful tool for pipeline management and for on-demand maintenance. Distributed strain
  monitoring also has the potential for detecting wall-thickness changes along the pipe, resulting from corrosion or abrasion.

In general, distributed strain/deformation and temperature sensing is a useful tool that ideally complements the current monitoring and inspection activities, allowing a more dense acquisition of operational and safety parameters. The measurements are performed at any point along the pipeline and not at specific positions only. Furthermore, the monitoring is continuous and does not interfere with the regular pipeline operation, contrary to, for example, pigging operations. The method can also be applied to non-piggable pipes.

Recent developments in distributed fibre-sensing technology allow the monitoring of 60 km of pipeline from a single instrument and of up to 300 km with the use of optical amplifiers. To achieve the above-mentioned goals and to take full advantage of the sensing technology described, however, it is fundamental to select and appropriately install adequate sensing cables, adapted to the specific sensing need. Although it is generally easier to install sensing cables during the pipeline construction phases, it is also possible to retrofit existing pipelines. In some cases it is even possible to use existing fibre-optic telecommunications lines installed along a pipeline for temperature monitoring and leakage detection.

6

Conclusions and Outlook

6.1 Conclusions

Structural health monitoring is not a new technology or trend. Since ancient times, engineers, architects and artisans have been keen on observing the behaviour of built structures to discover any sign of degradation and to extend their knowledge and improve the design of future structures. Ancient builders would observe and record crack patterns in stone and masonry bridges. Longer spans and more slender arches were constructed and sometimes failed during construction or after a short time (Levi and Salvadori, 1992). Those failures and their analysis have led to new insight and improved design of future structures. This continued struggle for improving our structures is driven by engineering curiosity, but also by economic considerations. Bridges and buildings must span larger valleys to foster new economic relations and faster movement of people and goods, and be constructed more cheaply and more durably. Now that in many western countries the main part of the transportation infrastructure is already built, we face new challenges related to maintaining this network in good condition and to serving its users in a dependable and secure way. We now realize that maintaining a large inventory of structures in good and safe conditions is even more challenging, and sometimes more expensive, than building them.

As for any engineering problem, obtaining reliable data is always the first and fundamental step towards finding a solution. Monitoring structures is our way to get quantitative data about our structures and help us in taking informed decisions about their health and destiny. However, the methods used for monitoring structures vary greatly: bridges are typically monitored only by periodic visual inspections, whereas dams and nuclear power plants are heavily instrumented and measured continuously. Advances in sensing and data processing technologies are extending the benefits of online monitoring to new classes of structures, such as bridges, buildings and historical monuments. Thanks to the progress in information technology and telecommunications, it is now possible to instrument a small bridge cost effectively and transmit the resulting data to a central database anywhere in the world.

In the last two decades, fibre-optic sensors have evolved from a laboratory curiosity to an industrial reality. Hundreds of projects have successfully demonstrated the reliability and durability of this technology in the most challenging applications. Fibre-optic sensors are no longer produced only by high-tech start-ups, but are now present in the catalogues of the most

Fibre Optic Methods for Structural Health Monitoring   B. Glišić and D. Inaudi
© 2007 John Wiley & Sons, Ltd

recognized suppliers of conventional instrumentation systems. Besides the benefits that can be obtained by replacing conventional sensors with their fibre-optic counterparts, it has also become apparent that new types of fibre-optic sensor can be used for entirely new monitoring tasks. In particular, long-gauge and distributed fibre-optic sensors have proven ideal for the monitoring of civil, geotechnical and oil and gas structures. These types of sensor enable the global monitoring of large structures with a reduced number of sensors and connections. This offers new possibilities, but also requires a new approach to the design and implementation of a monitoring system and to the processing and analysis of the resulting data. This book has presented the advantages and challenges related to the use of long-gauge and distributed fibre-optic sensors, guiding the reader in the selection of the most appropriate monitoring strategy and in the comprehension of the resulting data.

6.2 Outlook

In the near future we can expect that fibre-optic sensing technology will become a mainstream technology for SHM. The cost of these systems is declining, thanks to the phenomenal developments in the telecommunications industry, which uses many identical components. Fibre-optic sensors will increase significantly their market share in the global sensor market, no longer only a niche product. It is expected that new types of sensor will appear, but the main development will be in the consolidation, large-scale production and cost reduction of the exiting technologies presented in Chapter 2.

It is easy to predict that SHM in general will see a more widespread application to many types of structure that are currently not monitored or are only visually inspected. The driving forces in that direction will come from the necessity to optimize the maintenance investments, ensure the safety of aging structures, extend the lifetime of functionally obsolete structures, respond to new natural and man-made threats and increase our knowledge to improve the design of future structures. We are already witnessing the first examples of an entire 'fleet' of structures, in particular high-rise buildings, equipped with online monitoring systems.

As the hardware cost for a monitoring system continuously declines and makes it appealing for a larger number of structures, the costs of data analysis and interpretation will become the next retaining force to a more generalized use of this technology. It is important, therefore, that more research effort is devoted to the development of efficient and partially automated data analysis methodologies, translating the raw measurement data into high-level information that can be used for decision-making purposes.

On the international level, several efforts are currently devoted to creating guidelines and standards for SHM. The most notable initiative at the global level is supported by the International Society for Health Monitoring and Intelligent Infrastructures (www.ishmii.org), which also organizes international symposia on this topic. SHM is also taught as a specialization course in several universities worldwide; this will contribute to a better recognition of these methods and to more widespread use.

References

[1] Bernard, O. (2000) *Comportement à long terme des éléments de structures formés de bétons d'âges différents*, Ph.D. Thesis No 2283, EPFL, Lausanne, Switzerland.

[2] Brčić, V. (1989) *Strength of Materials (Otpornost Materijala)*, 6th edn, Građevinska knjiga, Belgrade, Serbia.

[3] Brönnimann, R., Nellen, P.M., Anderegg, P.G. *et al.* (1998) Packaging of fiber optic sensors for civil engineering applications. *Symposium DD, Reliability of Photonics Materials and Structures*, San Francisco, paper DD7.2.

[4] Bugaud, M., Ferdinand, P., Rougeault, S. *et al.* (2000) Health monitoring of composite plastic waterworks lock gates using in-fibre Bragg grating sensors. *Smart Materials and Structures*, 9 (3), 322–327.

[5] Casanova, N. (1998) Verformungsmessungen mit optischen Glasfasersensoren, *Messen in der Geotechnik 98*, IGB-TUBS Braunschweig – Institut für Grundbau und Bodenmechanik Fachseminar, Germany, pp. 1–14.

[6] CEB-FIP. (1990) *CEB-FIP Model Code*, fib, Lausanne, Switzerland.

[7] Cerulli, M., Inaudi, D., Posenato, D. and Glišić, B. (2003) EXPO 02, Piazza Pinocchio: monitoring visitor's live loads, *10th SPIE's Annual International Symposium on Smart Structures and Materials, Vol. 50577–48, San Diego, USA*.

[8] Christensen, R.M. (1991) *Mechanics of Composite Materials*, Krieger Publishing Company, Malabar, FL.

[9] Del Grosso, A. and Inaudi, D. (2004) European perspective on monitoring-based maintenance. *IABMAS 2004, International Association for Bridge Maintenance and Safety*, Kyoto, Japan (on conference CD).

[10] Del Grosso, A., Torre, A., Rosa, M., Lattuada, B. (2004) Application of SHM techniques in the restoration of historical buildings: the Royal Villa of Monza. *2nd European Conference on Health Monitoring, Munich, Germany*.

[11] Del Grosso, A., Torre, A., Inaudi, D. *et al.* (2005) Monitoring system for a cable-stayed bridge using static and dynamic fiber optic sensors. *2nd International Conference on Structural Health Monitoring of Intelligent Infrastructure (SHMII 2)*, Shenzhen, China. 415–420.

[12] Đurić, M. and Đurić-Perić, O. (1990) *Structural Analysis (Statika Konstrukcija)*, 4th edn, Građevinska Knjiga, Belgrade, Serbia.

[13] Ferdinand, P., Ferragu, O., Lechien, J.L. *et al.* (1994) Mine operating accurate stability control with optical fiber sensing and Bragg grating technology: the Brite–EURAM STABILOS project. *10th Optical Fibre Sensors Conference – OFS*, Glasgow, UK. 162–166.

[14] Ferdinand, P., Magne, S., Dewynter-Marty, V. *et al.* (1997) Application of Bragg grating sensors in Europe. *12th International Conference on OFS 1997 – Optical Fiber Sensors*, OSA Technical Digest Series, Vol. 16, Williamsburg, USA. 14–19.

[15] Frangopol, D.M., Estes, A.C., Augusti, G. and Ciampoli, M. (1998) Optimal bridge management based on lifetime reliability and life-cycle cost, *Short course on the Safety of Existing Bridges, ICOM&MCS*, EPFL, Lausanne, Switzerland, pp 112–120.

[16] Glišić, B. (2000) *Fiber Optic Sensor and Behaviour in Concrete at Early Age*, Ph.D. Thesis N° 2186, EPFL, Lausanne, Switzerland.

[17] Glišić, B. and Inaudi, D. (2002) Crack monitoring in concrete elements using long-gage fiber optic sensors. *First International Workshop on Structural Health Monitoring of Innovative Civil Engineering Structures, ISIS Canada, Winnipeg, Canada*. 227–236.

[18] Glišić, B. and Inaudi, D. (2003a) Components of structural monitoring process and selection of monitoring system, *PT 6th International Symposium on Field Measurements in GeoMechanics (FMGM 2003), Oslo, Norway*, 755–761.

[19] Glišić, B. and Inaudi, D. (2003b) Sensing tape for easy integration of optical fiber sensors in composite structures. *16th International Conference on Optical Fiber Sensors*, We 3-8, Nara, Japan.

[20] Glišić, B. and Inaudi, D. (2006) Finite element structural monitoring concept. *The 2nd fib Congress, Naples, Italy.* (on conference CD, paper ID 17-18, (#615)).

[21] Glišić, B., Inaudi, D., Kronenberg, P., Vurpillot, S. (1999) Dam monitoring using long SOFO sensor. *Hydropower into the Next Century*, Gmunden, Austria. 709–717.

[22] Glišić, B., Badoux, M., Jaccoud, J.-P. and Inaudi, D. (2000) Monitoring a subterranean structure with the SOFO System, *Tunnel Management International Magazine*, Vol. 2, issue 8, ITC Ltd, pp. 22–27.

[23] Glišić, B., Inaudi, D. and Vurpillot, S. (2002a) Whole lifespan monitoring of concrete bridges, *IABMAS'02, First International Conference on Bridge Maintenance, Safety and Management, Abstract on conference CD*, Barcelona, Spain, 487–488.

[24] Glišić, B., Inaudi, D., Nan, C. (2002b) Piles monitoring during the axial compression, pullout and flexure test using fiber optic sensors, *Transportation Research Record (TRR), Journal of TRB No. 1808 'Soil Mechanics 2002'*, paper N. 02-2701, Washington, DC, pp. 11–20.

[25] Glišić, B., Inaudi, D., Lau, J.M. *et al.* (2005) Long-term monitoring of high-rise buildings using long-gage fiber optic sensors. *7th International Conference on Multi-Purpose High-Rise Towers and Tall Buildings (IFHS2005)*, Dubai, United Arab Emirates. (on conference CD, paper #0416).

[26] Glišić, B., Posenato, D., Persson, F. *et al.* (2007) Integrity monitoring of old steel bridge using fiber optic distributed sensors based on Brillouin scattering. *The 3rd International Conference on Structural Health Monitoring of Intelligent Infrastructure, SHMII-3*, Vancouver, Canada. (on conference CD).

[27] Habel, W.R. and Hofmann, D. (1994) Determination of structural parameters concerning load capacity based on fiber Fabry–Perot-interferometers. *Proc. SPIE*, Vol. 2361, San Diego, CA. 176–179.

[28] Habel, W.R., Hillemeier, B., Jung, M. *et al.* (1998) Non-reactive measurement of mortar deformation at very early ages by means of embedded compliant fiber-optic micro strain gages, *12th Engineering Mechanics ASCE Conference, San Diego, USA. In Engineering Mechanics: A Force for the 21st Century ASCE*, Reston, USA, pp. 799–802.

[29] Hassan, M. (1994) *Critères découlant d'essais de charge pour l'évaluation du comportement des ponts en béton et pour le choix de la précontrainte*, Ph.D. Thesis No 1296, EPFL, Lausanne, Switzerland.

[30] Hughs, E.A. (2004) Live-load distribution factors for a prestressed concrete, spread box-girder bridge, *Master of Science in Civil Engineering thesis*, New Mexico State University Las Cruces, USA.

[31] Hughs, E.A. and Idriss, R.L. (2006) Live-load distribution factors for a prestressed concrete, spread box-girder bridge. *ASCE Journal of Bridge Engineering*, **11** (5), 573–581.

[32] Hughs, E.A., Liang, Z., Idriss, R.L., Newtson, C.M. (2005) *In-situ* modulus of elasticity for a high performance concrete bridge. *ACI Journal*, **102** (6), 458–468.

[33] Hurtig, E., Grosswig, S., Kühn, K., Schubart, P. (1996) Untersuchung von Sickerströmungen durch Dämme und Deiche mit Hilfe Faseroptischer Temperaturmessungen. *Mitteilungen Deutsche Geophysikalische Gesellschaft*, 118–121.

[34] Idriss, R.L. and Liang, Z. (2006) Monitoring an interstate highway bridge with a built-in fiber-optic sensing system. *IABMAS'06 – Third International Conference on Bridge Maintenance*, Safety and Management, Porto, Portugal. (on conference CD).

[35] Inaudi, D. (1997) *Field testing and application of fiber optic displacement sensors in civil structures, 12th International Conference on OFS 1997 – Optical Fiber Sensors, Williamsburg, OSA Technical Digest Series*, Vol. 16, pp. 596–599.

[36] Inaudi, D. and Glišić, B. (2002a) Development of a fiber optic interferometric inclinometer. *9th SPIE's Annual International Symposium on Smart Structures and Materials*, Vol. 4694-05, San Diego, USA. 36–42.

[37] Inaudi, D. and Glišić, B. (2002b) Long-Gage Sensor Topologies for Structural Monitoring. *The First fib Congress on Concrete Structures in the 21st Century*, Vol. 2, Session 15, Osaka, Japan. 15–16. (on conference CD).

[38] Inaudi, D. and Glišić, B. (2005) Development of distributed strain and temperature sensing cables. *17th International Conference on Optical Fiber Sensors*, Part I, Bruges, Belgium. 222–225.

[39] Inaudi, D. and Glišić, B. (2006) Reliability and field testing of distributed strain and temperature sensors. *13th SPIE's Annual International Symposium on Smart Structures and Materials*, Vol. SSM02, (Smart Sensor Monitoring Systems and Applications), Vol. 6167, San Diego, USA. 61671D-1–61671D-8.

[40] Inaudi, D., Casanova, N., Glišić, B. (2001) Long-Term deformation monitoring of historical constructions with fiber optic sensors. *3rd International Seminar on Structural Analysis of Historical Constructions*, Guimaraes, Portugal. 421–430.

[41] Inaudi, D., Glišić, B., Vurpillot, S. (2002) Database structures for the management of monitoring data. First International Workshop on Structural Health Monitoring of Innovative Civil Engineering Structures, ISIS Canada, Winnipeg, Manitoba, Canada. 85–94.

[42] Inaudi, D., Elamari, A., Pflug, L. *et al.* (1994) Low-coherence deformation sensors for the monitoring of civil engineering structures. *Sensors and Actuators A*, **44**, 125–130.

[43] Inaudi, D., Casanova, N., Kronenberg, P. *et al.* (1997) Embedded and surface mounted fiber optic sensors for civil structural monitoring. *Smart Structures and Materials Conference*, SPIE Vol. 3044, San Diego, USA. 236–243.

[44] Inaudi, D., Casanova, N., Steinmann, G. *et al.* (1998a) SOFO®: tunnel monitoring with fiber optic sensors. *Reducing Risk in Tunnel Design and Construction*, ITC, Basel, Switzerland. 25–36.

[45] Inaudi, D., Conte, J.P., Perregaux, N., Vurpillot, S. (1998b) Statistical analysis of under-sampled dynamic displacement measurement. *SPIE Symposium on Smart Structures and Materials*, Vol. 3325, San Diego, USA. 105–110.

[46] Inaudi, D., Kronenberg, P. and Vurpillot, S. *et al.* (1999a) Long-term Monitoring of a Concrete Bridge with 100+ Fiber Optic Long-gage Sensors, *SPIE's International Symposium on Nondestructive Evaluation Techniques for Aging Infrastructure & Manufacturing, Vol. 3587, Newport Beach, USA,* 50–59.

[47] Inaudi, D., Vurpillot, S., Martinola, G., Steinmann, G. (1999b) SOFO®: Structural Monitoring with Fiber Optic Sensors, *Fib Commission Meeting*, Vol. 40, No. 9, Institute or Structural Engineering, University of Applied Sciences, Vienna, Austria.

[48] Karashima, T. *et al.* (1990) Distributed temperature sensing using stimulated Brillouin scattering in optical silica fibers. *Optics Letters*, **15**, 1038.

[49] Keller, T. (2003) *Use of Fibre Reinforced Polymers in Bridge Construction*, IABSE-AIPC-IVBH, ETH Hönggerberg, Zurich, Switzerland.

[50] Kersey, A. (1997) Optical fiber sensors, S. Rastogi P. K. *Optical Measurement Techniques and Applications*, Artech House, 217–254.

[51] Levi, M. and Salvadori, M. (1992) *Why Buildings Fall Down*, W.W. Norton & Company, New York, USA.

[52] Liang, Z. (2004) Prestress losses and cambers of high performance concrete bridge girders, *Master of Science in Civil Engineering thesis*, New Mexico State University Las Cruces, USA.

[53] Measures, R.M. (2001) *Structural Monitoring with Fiber Optic Technology*, Academic Press, San Diego, USA.

[54] Muravljov, M. (1989) *Constructuion Materials (Gradjevinski Materijali)*, Naučna Knjiga and Faculty of Civil, Engineering, Belgrade, Serbia.

[55] Neville, A.M. (1975) Properties of Concrete, Pitman International.

[56] Niklès, M. *et al.* (1994) Simple distributed temperature sensor based on Brillouin gain spectrum analysis. 10th International Conference on Optical Fiber Sensors OFS 10, SPIE Vol. 2360, Glasgow, UK. 138–141.

[57] Niklès, M. *et al.* (1997) Brillouin gain spectrum characterization in single-mode optical fibers. *Journal of Lightwave Technology*, **15** (10), 1842–1851.

[58] Niklès, M., Vogel, B., Briffod, F. *et al.* (2004) Leakage detection using fiber optics distributed temperature monitoring. *11th SPIE Annual International Symposium on Smart Structures and Materials*, San Diego, USA. 18–25.

[59] Radojicic, A., Bailey, S. and Brühwiler, E. (1999) Consideration of the serviceability limit state in a time dependant probabilistic cost model, *Application of Statistics and Probability*, Balkema, Rotterdam, Netherlands, Vol.2, 605–612.

[60] Talbot, M., Laflamme, J.F., Glisic, B. (2007) Strees measurements in the main cable of a suspension bridge under dead and traffic loads, *EVACES'07 - Experimatal Vibration Analysis for Civil Engineering Structures*, Porto, Portugal. (Paper #138, on conference CD)

[61] Thévenaz, L., Niklès, M., Fellay, A. *et al.* (1998) Truly distributed strain and temperature sensing using embedded optical fibers. *SPIE Conference on Smart Structures and Materials*, Vol. 3330, San Diego, USA. 301–314.

[62] Timoshenko, S.P. and Goodier, J.N. (1970) *Theory of Elasticity*, McGraw-Hill International.

[63] Udd, E., Nelson, D., Lawrence, C., Ferguson, B. (1996) Three axis and temperature fiber optic grating sensor. Smart Structures and Materials 1996, SPIE Vol. 2718, San Diego, USA. 104–109.

[64] Vurpillot, S. (1999) *Analyse automatisée des systèmes de mesure de déformation pour l'auscultation des structures, Ph.D. Thesis No 1982, EPFL, Lausanne, Switzerland.*

Index

Some of the words in the index occur in several pages, and the authors believe that the listing of all the corresponding page numbers may not be beneficial to the readers. For such words, the authors decided to highlight in bold the most important pages (if possible to discriminate them) or simply to list only the most important ones highlighted in bold and to omit other occurrences. Thus, if the number after a word is either a normal character or mixed bold/normal character, then the list is exhaustive (e.g. the words 'Activity' and 'Analysis' respectively); if the full list of pages after a word is only presented in bold, then it is not exhaustive, but only the most important occurrences are listed, while the others are omitted (e.g. the word 'Bridge').

Fibre Optic Methods for Structural Health Monitoring   B. Glišić and D. Inaudi
© 2007 John Wiley & Sons, Ltd